THE APPROACH TO INDUSTRIAL CHANGE IN BRITAIN AND GERMANY

THE APPROACH TO INDUSTRIAL CHANGE IN BRITAIN AND GERMANY

A Comparative Study of Workplace Industrial Relations and Manpower Policies in British and West German Enterprises

By
Eric Jacobs
Stanley Orwell
Peter Paterson
Friedrich Weltz

The research for this publication was financed by the
ANGLO-GERMAN FOUNDATION FOR THE STUDY OF INDUSTRIAL SOCIETY
as part of its programme of study of social and economic policy issues

St. Stephen's House, Victoria Embankment, London SW1A 2LA
5300 Bonn 2, Jean-Paul-Strasse 12

The opinions expressed in this report are entirely those of the authors.

Series B 0179

ISBN 0 905492 06 4

Price: £10.00

PRINTED IN GREAT BRITAIN BY
BILLING & SONS LIMITED, GUILDFORD, LONDON AND WORCESTER

THE APPROACH TO INDUSTRIAL CHANGE IN BRITAIN AND GERMANY

A Study by

Social Policy Research Limited, London

and

Friedrich Weltz, Munich

for

The Anglo-German Foundation
for the Study of Industrial Society

Social Policy Research Limited is a charitable company limited by guarantee, which carries out studies into the policies of government and industry. SPR draws on the resources of the MIL Research Group and on consultants from industry, the professions and universities.

Friedrich Weltz is an industrial sociologist, mainly engaged in consultancy and research assignments for government, foundations and private industry.

Contents

Contents

Acknowledgements

We are grateful to the companies and unions whose collaboration made this study possible, and to the members of management, union full-time and lay officials and other representatives of the workforce on whose experience and attitudes our report is based.

However, it should be emphasised that while companies and unions provide the material, the authors take entire responsibility for its evaluation.

This study, carried out mainly in 1976, is based on an earlier appraisal on problem areas and industries in a state of change. As part of this appraisal, an Anglo-German seminar was held in London in May 1975. A number of participants to the seminar contributed background papers (now lodged with the Foundation).

We acknowledge their contribution, while making it clear that they took part in the seminar as individuals (occupations at the time are given overleaf only to show fields of activity) and that they are not responsible for our use of the material.

Contributors to the 1975 seminar:

S. D. Anderman	Senior Lecturer in Law, Warwick University
Fritz Bohle	Institut for Sozialwissenschaftliche Forschung
John Fyfe	Economic Adviser, Manpower Services Commission
K. Graham, OBE	Secretary, Organisation and Industrial Relations Department, Trades Union Congress
F. C. Hayes	Deputy Chief Executive (Industry) Training Services Agency
Gerhard Leminsky	Wirtschafts-und Sozialwissenschaftliches Institut des DGB

Helmut Minta	Vice-President, Bundesanstalt für Arbeit
E. J. Robertson	Director—Industrial Relations, Confederation of British Industry
Werner Sorgel	Social research and consultancy, Munich
Dorothy Wedderburn	Director, Industrial Sociology Unit, Imperial College of Science and Technology

Members of the study team in the two countries:

In Britain

John Dickson, lecturer in management studies
John Endacott, management consultant
Roger Fox, lecturer in industrial sociology

In Germany

Gert Schmidt, lecturer in industrial sociology
Rainer Erd, industrial sociologist
Walther Müller-Jentsch, industrial sociologist

Objects

The study deals with employment problems created by rationalisation and re-organisation in British and West German enterprises.

The aim of the study was to discover how management and representatives of the workforce in the two countries respond to key employment problems at a time of change. We attempt to explain this response in terms of the principles governing employment policy and industrial relations at enterprise and plant level. This in turn has led us to examine the way in which conflict is managed (or avoided).

Britain and Germany have similar industrial structures and broadly similar patterns of employment, and they also face similar problems in the restructuring of industry as a result of technological development and economic pressures. However, these problems are set against very different backgrounds of economics, labour law and trade union organisation. There is, therefore, a danger of comparing what appears to be like with like. Nothing productive could emerge from comparing blow by blow, and out of context, the introduction, say, of continuous casting in a British as against a German steel works.

The project began with an appraisal of the problem areas and industries in which a comparative study was likely to be most productive (reported to the Foundation in July 1975).

This appraisal pointed to the need for research at enterprise level. The employment policies of enterprises are affected by such diverse forces as the political and economic state of the country, competition between concerns, the state of the labour market and negotiations at national level between employers and unions. None the less, a study of these superior forces cannot tell us how they interact at enterprise level which is where changes such as major rationalisation actually occur.

The earlier appraisal also pointed to the value of studying

enterprises' manpower policies as a key to larger effects on labour markets.

Many earlier studies have treated employment policies as an outcome of economic or technological change. This study differs in taking enterprise employment policy as its main focus not as the outcome of something else. The result is not just a subtle difference in viewpoint, but an opportunity to examine in depth the interaction of employment policies with industrial relations.

Our study concentrated on large-scale enterprises which collectively account for a large part of total employment and are of special importance for the national scene in having the power to set the pattern for future developments.

The study was also confined to large changes such as the re-organisation of plants, movement of production from one site to another, the introduction of computer technology and large reductions in the workforce. In what we believe to be the first substantial comparative study of its kind in Britain and Germany, it seemed appropriate to concentrate on the larger changes.

This concentration on large change was made in full recognition that at the same time there is a continuous flow of small changes each affecting a few jobs or aspects of a work routine. These small changes taken together change the character of industrial work less dramatically than large changes we have described but no less fundamentally.

We sought for "successful" examples of change not to produce copy-book models but to make it possible to examine the complete process of how changes are followed through and conflicts resolved. Some of the studies in both countries dealt with recent history, others followed through change as it occurred. In the latter cases, there was no certainty that the outcome would be "successful".

Approach

We sought, with the aid of employers and unions, parallel examples in Britain and Germany which filled our requirements of large enterprises which had gone through major changes recently and "successfully".

We chose sections of the coal and steel industry in both countries as examples of long-term fundamental change and for the added reason that, in Germany, they represent co-determination (equal representation of workers and employers on the supervisory board) which cannot be excluded from any study of change. In retail distribution and in the paper industry we again found closely parallelled examples. However, when we move to the motor industry where important changes have taken place in Germany, which could not be excluded even from a small number of case studies, we found no direct parallel in Britain. Consequently, in Britain we had to take other examples in the field of engineering. Finally, to indicate the effects of computer technology, we took newspapers in Britain and Germany as well as an insurance company in Germany, which revealed widely contrasting responses in different industrial environments.

In all, seven enterprises were selected in each country.

In each enterprise we began by approaching general management, both to enlist their cooperation and to discover the processes of economic or technological change which had led to effects on the workforce. Typically, we then moved on to the personnel department, both to provide data and to guide us on who else to see in management and as representatives of the workforce. In most cases, we also made a parallel approach through the unions involved, following a route from the national to local full-time officials and then to lay officials in the enterprise (often already identified by personnel department).

Each case study involved between 10 and 30 interviews supplemented by appropriate documentation, company reports, plans, etc. Both companies and unions generously gave access to this data, and it forms an important part of our evidence.

We originally intended to carry out the studies at enterprise level since our aim was to look at the approach to major problems. However, in the event this proved too restricting an aim. On the one hand, most of the studies had to be pursued to the level of a site or sites—such as a factory or retail outlet—since this was usually the level at which actual change took place. On the other hand, some studies had also to be pursued beyond the enterprise, to the national industrial organisations, since in those cases change could only be understood in a wider context.

Case studies were the only feasible method of carrying out this investigation among large industries within the resources available to us, or indeed which would be reasonable to devote to a study of this kind. They provide examples of policies, industrial relations, resolution of conflict, etc., but their interpretation depends on a wider industrial context. Here lies the importance of the earlier appraisal which we carried out at the first stage of the project, including the preparation of background papers on union organisation, labour law and the labour markets and industrial society in the two countries. The report which follows has drawn widely on this material as a framework for the case studies.

Outline

The report is divided into three parts—a chapter on industrial change in Britain, a chapter on industrial change in the Federal Republic, and a comparative chapter which, while not drawing hard and fast conclusions on the parallels provided by case studies in each country, does suggest some areas where further exploration and discussion would be worthwhile.

The British Report

Section I deals with the historical background of the British industrial relations system, emphasising that its dominant style is of an adversary nature. It discusses the origins and the legal position of the British trade union movement and points out that the multiplicity of trade unions in many industries is a complicating factor in the management of change. The importance of shop stewards as the key agents in the bargaining of change is stressed. The organisation of employers' associations is also considered, with their close resemblance to the trade union structure in numbers and fragmentation. Finally, this section deals with the role of government in industrial relations. State intervention, it says, has taken three major forms: in improving the conditions of employment of the individual workers; in regulating the behaviour of union and employer organisations; and in improving the operation of the labour market to raise the performance of the economy as a whole.

Section II summarises the case studies conducted at seven enterprises, and which are dealt with more fully in the lengthier, and separately published, British report.

Section III deals with employment policy. It draws attention to the "preferred option" of British management in reducing costs to dismiss a proportion of their employees. One reason for this is the concentration of legislation designed to protect the individual worker at the point of dismissal.

Another is the trade union structure which can set up demarcation barriers inhibiting flexible transfers between job classifications, and in some instances the establishment of viable internal labour markets within the enterprise.

Section IV examines industrial relations as an influential factor in the handling of rationalisation and re-organisation within the enterprise. In the absence of a uniform system for dealing with change, it notes that management tends to invent a tactical approach according to the institutional arrangements existing in the plant affected by change. It observes that managements frequently prepare their plans secretly, making it difficult for trade unions to influence them in the formative stage, and often presenting them with a planning fait accompli. The characteristic most easily identified in a number of the case studies was the unpredictability of union-management relations where more than one union was recognised, and where there was no permanent joint negotiating machinery enjoying the confidence of both sides. Change was most easily achieved where there was no significant union presence, or where a close relationship had been built up by management with a single, dominant trade union.

Section V recognises that a new framework is emerging in Britain, based on the philosophy that by giving trade unions as of right a larger element of participation and involvement within the system, they in turn will demonstrate a greater sense of commitment to, and responsibility for, the running of industry.

The German Report

Section I examines the economic development of the Federal Republic, concluding that growth has been assisted by the remarkable stability of industrial relations both at industry level and at the level of the individual enterprise. Rationalisation has been regarded by the unions as a necessary and unavoidable part of economic development, and they have not tried to impede the process of technical change.

Section II recapitulates, in summary, the seven case studies undertaken for the longer German report.

Section III is concerned with manpower policy. It finds

that manpower policy is not regarded as the most important factor in the initial causes of innovation, the case for change being seen primarily as an economic or technical problem. Since the 1973 oil crisis, German industry has turned from "expansive" to "intensive" innovation, with the number of job opportunities declining, and a greater emphasis placed on retraining, redeployment or wastage. Faced with this new situation, managements have developed "silent" employment policy, a low-key approach that brings flexibility as a corrective to the legal framework of "official" employment policy. It is a direction that needs the close cooperation of works councils.

Section IV is concerned with industrial relations. It finds that the planning and implementation of change in the case study enterprises was, with the major exception of the printing industry, free of conflict and regarded by neither side as controversial. Works councils were informed of proposals at an early stage, and the conditions they sometimes laid down were accepted by management. These included the avoidance of dismissals, the retention of their former pay rates for a limited period by workers downgraded as a result of change and the presence of works councils representatives at meetings where staff transfers were being considered.

Section V discusses a developing style of industrial relations in West Germany which is described as "cooperative conflict resolution". This style is characterised by the concern of management and works councils to avoid open confrontation; a willingness to seek compromises; the renunciation of the chance to exploit legal rights to the extreme; acknowledging that the interests of the two sides may well diverge; and the encouragement of informal agreements in line with the requirements of the enterprise without reaching the point at which the letter of the law must be invoked. This section also looks at other styles which exist in the Federal Republic, including open confrontation and the suppression of conflict. Cooperative conflict resolution, it states, represents a middle way between them.

Section VI examines the influence of external factors on the conduct of change within the enterprise. These include the Law on Works Councils and Co-determination, nationally negotiated collective agreements and the style of employer-union relationships at national level. The report identifies a climate of predictability that allows long-term planning and innovation to be introduced with comparative ease.

Britain and West Germany Compared

Section I sketches in the background of industrial change in both countries, finding that while industrial relations in Britain are conducted in a glare of publicity, the German system conveys the impression of a well-ordered, smooth-running system. The prevailing style in the Federal Republic is essentially cooperative, while in Britain it takes an adversary form—though cooperation exists in Britain just as contention exists in Germany.

Section II is concerned with aims and methods. The overriding objective of change in each country is to achieve a reduction in labour costs, although the means chosen differed. In Britain, the solution almost invariably chosen was dismissal: in Germany, early retirement, a halt to recruitment and the redeployment of manpower within the enterprise were used. For German management, stability of employment is a conscious objective—for British management it is desirable but not a priority. This has tended to inculcate a more acute sense of anxiety towards change among British workers. In the absence of multi-unionism, institutional conflicts within the enterprise do not arise as they do among competing unions in Britain.

Section III looks at the institutional and legal framework. Its main conclusion is that post-war legislation in Germany has concentrated heavily on establishing and building up enterprise level cooperative institutions while retaining the autonomy of collective bargaining. The British experience has featured attempts to reform the institutions and the laying down of minimum legal standards regardless of the institutional arrangements. The main thrust of legislation has been towards regulating and improving workers' rights at the point at which their jobs are at risk. In Britain, trade unions form a virtual single channel of representation for workers. In Germany a dual system exists, with trade unions operating largely outside the workplace, and legally based works councils providing an enterprise-based negotiating partner for management on matters affecting the enterprise.

Section IV shows how entirely different approaches to change are adopted in each country because of the contrasting institutional arrangements. It states that the German system does not lack conflicts of interest, and that the performance of works councils is not always uniform, some adopting an abrasive attitude towards employers and others sometimes over-identifying their interests with those of the employer. A middle way has developed with the style of cooperative conflict resolution.

Section V examines the costs of industrial change on the community and on the worker. In both countries heavy costs are transferred to the community, either by management dismissing workers, as in Britain, or by the restriction of job opportunities implicit in the intensive internal labour markets operated by large German enterprises. The report finds that the greatest difference between the two countries is in the uniformity of the German institutional framework, as against the bewildering variety of practices in Britain.

Section VI sums up our findings for future consideration, suggesting key elements in the institutional experience of the two countries which may be useful to the other and proposing measures for enterprise employment policy.

CHANGE IN BRITAIN

I

THE BACKGROUND

1. The aim of this chapter is to explore the performance of a number of British enterprises in handling the problems of industrial change. It is primarily based upon case studies undertaken in seven concerns which had gone through, or were in the process of implementing, schemes of rationalisation and re-organisation. We think it necessary in giving a coherent account of British experience in this field to devote considerable attention to the recent and historic circumstances which have produced the intricate structures within which change necessarily occurs. Our method, therefore, is first to describe this background, next to give a brief account of each of the case studies, and then to explore the implications of both for employment policy and industrial relations.

2. The dominant style of the British industrial relations system is its adversary nature. Of course, the seeds of conflict and dispute exist in every industrial society and they are from time to time expressed in strikes and lockouts wherever such free expression is permitted. Classically, and still most commonly today, disagreement occurs over the division of the product of industry between profits and wages. But in Britain the adversary style, while it embraces this kind of argument, also goes much further. It is the favoured means of addressing almost every kind of industrial problem that crops up. The great majority of issues are settled amicably enough, but conflict potentially escalating into strike action is inherent in the adversary approach. It is a possibility that can never be remote from the mind of a management contemplating change or innovation.

Perhaps the adversary style is characteristically British,

rather than merely peculiar to British industry. It is typical of the legal system with its advocates for prosecution and defence, and of the parliamentary form of government, in which the duty of the parties not represented in the Government is conceived to be simply to oppose—hence their collective title, the Opposition. Indeed, the analogy between Parliament and industry has been elevated almost into a theory of industrial democracy. In a well-known essay, Professor Hugh Clegg has cast the trade unions in the role of "permanent Opposition", like a political party that can never form a Government. (*Industrial Democracy and Nationalisation*, Oxford 1951.)

3. The key component in the system of adversary industrial relations is undoubtedly the trade unions. It is on them that, above all, responsibility for representing employees rests at all points in the economy, from negotiations with government on national policies to the details of industrial change at enterprise level which are the concern of this report. Their structure and methods are therefore vitally important to an understanding of the processes of industrial change.

The British trade union movement is not divided on religious or political grounds (as in France and Belgium) nor on occupational grounds (as in Sweden). The vast majority of all British trade unionists belong to one or another of the 113 unions affiliated to the Trades Union Congress, a national union centre that has already more than 100 years of activity behind it.

In this sense—that is, membership of the TUC—the British trade union movement can be said to be united and at one. All skills, all grades of employee and all shades of opinion—from Maoist to Catholic—can be and are accommodated within it. But the appearance of unity is deceptive. For the common fact of TUC membership conceals a very real diversity—particularly of union organisation, structure and constitution. The TUC retain its unions in membership only by exercising a very light weight of control over them.

4. The diversity of British trade unions lies in their origins. They emerged one by one in order to meet the needs of the groups of workers who were to become their members.

Roughly, one may speak of craft unions (like the Engineers), unskilled general unions (like the General and Municipal Workers' Union), industrial unions (like the Miners) and, more recently, specifically white-collar and professional unions (such as the Association of Scientific, Technical and Management Staffs). But each description needs to be modified. The Amalgamated Union of Engineering Workers, for example, no longer confines its membership to apprenticed craftsmen but recruits unskilled labour as well. At the same time a general union like the Transport and General Workers will happily compete with the Engineers for skilled members. Nor is there such a thing in Britain as a genuine industrial union. The National Union of Mineworkers comes very near to it, but even in their own industry there are two other unions whose views have to be reckoned with. Although there have been many mergers between unions in the last decade, this has done little to make the overall structure more rational. Competition between unions for members—and so for income, power and status—is fierce, and with the opening up of new fields for recruitment among hitherto unorganised white-collar groups, there are signs that the competition is becoming yet more intense, and with the further complication of the appearance of new unions, outside the TUC, to cater for the new trade unionists.

5. If British unions defy precise classification, this is more than a matter of regret to the tidy-minded. It means that industrial problems often have to be sorted out, not just between a management and a single union, but between management and several unions. And each union may be very different: in its aims and interests, its rules and procedures, and in the degree of control it exercises over its members. The multiplicity of unions is not necessarily bad, but it is certainly a complication. Like the ever present possibility of a strike inherent in the adversary system, it is a factor that management has to bear constantly in mind when it is embarking on change. It has to be acknowledged that the TUC has, organisationally, done much at national level—particularly through its comparatively recent industry committees—to diminish the consequences of multi-unionism. But there is a long way to go before equivalent bodies at local

or plant level are working as effectively. Such a system may, in our view, be one of the most useful by-products of an advance towards industrial democracy—a move which, following the Bullock Report, is firmly on the political agenda in Britain.

6. The unions were emancipated just over 100 years ago in an oddly negative form, which they largely retain today. They were regarded as organisations whose aims were to restrain free trade, like any commercial monopoly. Nevertheless, they were too powerful simply to abolish. So they were relieved of the penalties that fell on other restrainers of trade. They became a sort of licensed conspiracy; tolerated, if not approved. Thereafter and for most of the next century, any handful of men who got together to pursue an industrial objective enjoyed the immunities of a trade union. Providing their aims were purely industrial (for instance, a wage increase), they could not be taken to court for any losses they might cause to an employer. At the very formative stage of trade union development the law, in effect, turned its back on them, and they grew and developed as they pleased. This has bred in the unions a very distinctive feature: a sense of private sovereignty. They alone, British unions devoutly believe, are entitled to settle their internal affairs. The unions dislike interference of any kind, whether from other unions, the TUC or from government. So, when their own apparent inability to control their members caused governments to try from the late 1960s to intervene to set standards for union behaviour, they proved highly resistant to the experiment. The law had neglected them for so long that it seemed almost too late for it to become involved.

This sense of sovereignty has had another important consequence: an unwillingness on the unions' part to tolerate any complementary or alternative form of worker representation, such as a statutorily imposed works council system. No uniform representative system has, therefore, been able to develop across industry, and where no unions exist, workers are likely to be represented by management-approved bodies or staff councils or "house" unions. In the main, these have proved ineffectual, though there are rare exceptions.

7. One further feature of this system of representation needs to be emphasised at this stage—the erosion of union central authority. This was acknowledged by the Royal Commission on Trade Unions and Employers' Associations under Lord Donovan (1965-68) which described two distinct systems of industrial relations: the formal and the informal. The formal system consists of the bargaining structures established by employers and trade unions at national and district level, with elaborate rules and procedures for the settlement of disputes and the fixing of wages and conditions of employment. Alongside is the informal system in the plants, run by shop stewards (of whom there are estimated to be some 300,000). Stewards from different unions may combine into a joint committee for negotiating purposes, but there is no legal requirement for them to do so, and even where such bodies do exist there is no guarantee that individual shop stewards will not take independent action. The main impetus behind this dispersal of authority from union headquarters to the shop stewards has been the development of plant level wage bargaining since the war. But for our purpose, its importance is that power has been acquired by shop stewards who tend to act very independently, and are usually key agents of bargaining over industrial change within the enterprise.

8. Worker representation, as far as it affects the approach to change, can be described like this: unions claims the right to act as the channel through which consultation and negotiation with management takes place; union organisation at plant level varies widely, according to the number of unions with members in the enterprise, and it also depends heavily on the work of voluntary, unpaid shop stewards; there is no set of uniform rules and procedures for handling change across the whole of industry; representatives of different unions within an enterprise may combine to alleviate some of the problems of multi-unionism, but it is not mandatory for them to do so, and unions do not thereby surrender their right to autonomous action; in the absence of a union, workers are likely to be without any form of effective representation.

9. We have described one-half of the adversary relationship; we now briefly describe the other half—the employers. In Britain, employer organisations arose in response to the development of the unions, which they in some ways resemble, especially in their structure. Like the unions, they are fragmented; in fact, there are even more employer organisations than there are unions. And, like the unions, they deploy a weak central authority. The effective use of joint resources to combat union action—parallel to a union's strike fund—though not entirely—is almost unknown. Individual firms rely on their associations at most to negotiate with the unions minimum terms and conditions for their industry and to provide advice and services on, for example, legal and productivity problems.

10. The identity between the Employers' Organisation and the union should not be overstressed. Many such organisations undertake a far wider range of functions than a union, representing their members not only in labour relations matters, but also on many technical, trade and commercial questions. This is particularly so in the case of the Confederation of British Industry—often, rather misleadingly, known as the employers' TUC. For all these reasons the responsibility for handling the problems of change with which we are concerned in this report devolves almost entirely on the company itself. Within companies no standard system for dealing with employee relations exists. There are still some who refuse to recognise unions at all, although this is increasingly rare, and more managements find it convenient to try to resolve their difficulties through collective bargaining. But the typical large company now has a board member with special responsibility for labour affairs, a developed personnel department and a commitment to recognise and bargain with a number of unions representing various grades of employee.

11. In this context, we would emphasise only one further aspect of employer organisation. Given the fragmented structure on both sides of the adversary relationship, it appears more difficult for employers to resist union claims than it was in the past. If one union, representing a fraction of employees in a critical area of the production process, puts

forward a demand, a firm's ability to resist is weakened by the fact that it is likely to find itself continuing to pay those employees—often the majority—who stay at work while the others threaten to strike. And at the same time the enterprise can expect no support from other firms in the same trade who may indeed stand to gain as a result of any production losses sustained by a company in dispute with its employees. In short, within the adversary system as it has evolved in Britain, the employers are increasingly the weaker partner in the relationship. They no longer enjoy, in dispute with unions, the superiority in "firepower" that was once automatically associated with ownership of capital. Part of the reason for this lies in greater national and international competition, and in the vulnerability of companies to pressure on cash flow caused by an interruption of output. But part of the responsibility for altering the traditional power balance also lies with the State. In Britain, the cost of financing strikes is more and more borne by government—in practice if not in intention—through tax rebates and social security payments to strikers' families; but no such public money is available to help out companies affected by a strike.

12. This funding of strikes is only one of many examples of the way in which the State has become entwined with the industrial relations system in Britain. Industrial relations can be the concern of any national government when, for example, a national strike occurs in a vital industry or service. But in Britain, the State has become more regularly and more closely involved than in most countries because successive governments have perceived defects in the workings of the adversary style—for instance, a high unofficial strike rate and wage-induced inflation—and have over the past ten to fifteen years attempted to repair them.

13. State intervention has taken three major forms. First, it has tried to improve certain conditions of employment of the individual worker, on which collective bargaining has made little impact. Most notably, the Contracts of Employment and Redundancy Payments Acts were introduced in the 1960s to provide minimum periods of notice and financial compensation when a worker was dismissed for economic reasons. In 1971, the Industrial Relations Act introduced the

concept of "unfair" dismissal, under which an employee had the right to appeal to a tribunal for financial compensation on the grounds that he, or she, had been wrongly dismissed. These measures have since been supplemented by the Trade Union and Labour Relations and Employment Protection Acts, which have further improved workers' conditions in, for example, giving a right to time off to seek new work, and to a three-month warning period to unions, in both cases when there are impending redundancies. There are other important features of these Acts to which we shall return, but here we would simply note that the concentration of legislation has been on the point where workers' jobs are at risk. With such exceptions as health and safety, maternity rights for women, and statutes that make discrimination at work for reasons of sex or race unlawful, the employment conditions of people who are not threatened by dismissal have been left almost entirely to be determined by collective bargaining within the enterprise.

14. The second major thrust of State intervention has been towards regulating the behaviour of union and employer organisations, both internally and in their relations with each other, with the common aim of improving the performance of the collective bargaining system. But two false starts were made, one by Labour in 1969 and a second by the Conservatives in 1971. Both attempted to initiate legislation which ultimately failed because the unions resisted what they interpreted in both instances to be an unacceptable interference in their internal affairs and an erosion of their traditional rights and freedoms. The Conservative legislation was repealed and replaced by the Trade Union and Labour Relations and Employment Protection Acts, which put a very different face on State intervention. Instead of laying down rules and procedures for the unions to follow, these Acts offered the unions a whole range of improved rights and opportunities. The spread of the closed shop form of union organisation was given specific legal backing; protection was afforded to workers carrying out union duties; new opportunities for union recruitment and organisation and subsequent recognition by employers were established; and unions were to be afforded access to company information

needed to support collective bargaining. Accompanying this new wave of union rights came proposals for unions to sit on the governing bodies of private pension schemes, to appoint safety officers in factories and offices, and, with the Bullock Report, to appoint nominees to the boards of directors of all companies employing more than 2,000 people. After some years of experiment and trial and error, the new direction of State intervention was now clear: a single-channel system of worker representation was to be recognised throughout the economy, and that exclusive channel was to be the trade union movement.

15. The third direction of government policy has been economic. A series of steps has been taken to improve the operation of the labour market as part of a more general effort to raise the performance of the economy as a whole. Some of the legislation already mentioned shared this economic purpose. For example, the Redundancy Payments Act was intended not only to give the worker rights, but to make him more willing to change his job to suit new employment requirements. The most important single piece of "economic" labour legislation was the Industrial Training Act, which set up a series of industry-level boards, jointly run by unions and employers, to expand training opportunities and standards. The importance attached by governments to the economic aspects of manpower has now been recognised in the establishment of the Manpower Services Commission to oversee both training and job placement.

16. The industrial relations system in Britain, which is the vehicle required to cope with the social aspects of change, can now be summed up as follows: the style of worker-employer relationships is adversary; the structure of employer and union organisations is fragmented, no more than a fragile authority being exerted by each over constituent members; government legislation has aimed at improving the performance of the system by enlarging individual rights (primarily at the point where jobs are at risk), widening the collective rights of trade unions, and making the labour market work more efficiently; and the drift of all this is towards making the trade unions the single channel of representation for all employees. It should be noted, however,

that many of these developments were still evolving when the events described in the case studies—which we now summarise—occurred.

II

THE CASE STUDIES

17. The enterprise involved in the studies were:

sectors of two nationalised industries:
the Scottish Division of the British Steel Corporation and the North Staffordshire area of the National Coal Board

a small industrial conglomerate:
Harris & Sheldon Group Limited

a department store group:
Debenhams Limited

a newspaper:
The Financial Times

and a large paper-making company:
Bowater's

A unitary engineering concern was also examined but, at the request of its management, is not identified; we shall refer to it as Company Y.

Bowater

18. Several developments came together to threaten the viability of this company's paper-making activities. First there was heavy competition from rivals in the United Kingdom, such as the Reed Paper Group. Then competition was intensified from abroad, partly by the decision of Scandinavian pulp producers to move into paper-making, and partly by the reduction of European tariffs through the EEC and EFTA. This was particularly serious since it was

accompanied by a sharp rise in the price of wood-pulp, the basic raw material—much of which came from Scandinavia—to a point where UK producers like Bowater found it difficult to compete on finished paper prices. In any event, worldwide overcapacity in paper-making had been created by the early 1970s, so that Bowater's problems were naturally intensified when the economic recession struck in the wake of the oil crisis. Against this background, Bowater decided to move out of the production of the cheaper grades of paper and of the top quality grades in order to concentrate on wood-free papers (a technical term meaning that the paper is made from wood that is chemically treated rather than ground up) with the raw material supplied from domestic rather than imported sources. The result of this series of pressures and events was that over a period of time the company had to concentrate its production on fewer machines and production sites.

19. As early as 1971 Bowater closed six of its twenty paper-making machines, which resulted in 1,600 redundancies, and three years later the company decided to close down one mill completely and to transfer part of its production to another. The mill that was closed was the Thames at Northfleet in Kent, and the one to be expanded was the Kemsley Mill twenty miles away. Briefly, the company's aim was to rebuild an unused machine for wood-free paper production at Kemsley, to close Thames at a cost of 570 blue-collar jobs, and to encourage as many as possible of those displaced workers to transfer to Kemsley to fill 403 new jobs. Once the basic decision had been taken, management prepared a highly detailed plan which was named "Operation Penthouse" (so called because early planning took place at a penthouse-type office in the company's engineering division at Gravesend). The script for Operation Penthouse was a 31-step programme detailing every stage on the way to completion of the project, beginning with a submission to the Bowater board of directors, going on to describe necessary changes in management structure and, most crucially, outlining the tactics for winning the cooperation of the employees and their unions. Nothing that could be foreseen was left to chance.

20. The management anticipated that the closure of Thames Mill, the transfer of the workers and the rebuilding of the wood-free machine at Kemsley would take 16 months to complete. The calculation included the time necessary for elaborate consultations with the unions and to give the workers, whose lives would obviously be disrupted, time to make arrangements for the move.

21. The two principal unions at Thames Mill and Kemsley were the Society of Graphical and Allied Trades (SOGAT), whose members worked in the paper-making plant, and the Transport and General Workers' Union (TGWU), which organised workers in the goods-handling and ancillary departments. For both these groups of workers, union membership was almost total. Among staff employees unionisation was lower, although there was a substantial membership in both the TGWU white-collar section (ACTS) and the ASTMS. Among the blue-collar unions relations were harmonious; SOGAT shop stewards at Thames Mill joined other unions on a joint committee, which they frequently declined to do in other parts of the industry.

22. A carefully gauged order of priorities was drawn up for informing all concerned in the operation. It began with a secret meeting with top SOGAT officials, went on to the national and local officials of the other unions, the firm's customers, the Department of Employment, which might be involved in finding jobs for workers who did not transfer, and with the three local councils and the MPs for the area. After a final briefing for managements at all the company's mills came "A-Day"—28 March 1973. On that day, starting at 8 a.m. an intensive series of meetings monitored from a control centre at Northfleet broke the news to all personnel, starting with departmental managers, continuing with shop stewards and supervisors, and finishing at 11.30 p.m. when workers on the last shift were informed. Management took great pains to preserve the secrecy of their plans until they were ready to disclose them.

23. Why did the Bowater management adopt a policy of secrecy in formulating their plan? First there was the commercial factor: the company wished to retain the con-

fidence of its customers and to avoid a premature halt in production at Thames Mill before the new installation at Kemsley was completed. But perhaps more important was uncertainty as to how the unions and their members would react. On the union problem it was seen as an important priority to "get them on our side" nationally. Psychologically, the message management wished to get across at all levels was that the emphasis of the project was on the transfer of work rather than the closure of a mill. After an initial display of resistance, the unions cooperated fully in the scheme—indeed, the most vociferous public protest over the company's action came from the businessmen of the local Chamber of Trade and from the town of Sittingbourne, heavily committed to the paper-making industry and still feeling the shock of earlier closures. After a brief flurry in the local press, the campaign of resistance petered out.

24. All along, the company's underlying fear was of active opposition to the closure by the workers at the Thames Mill fearing the loss of their jobs. For this reason, the plan of action involved offering all the blue-collar workers alternative employment at the new location, even though it would have been an embarrassment to Bowater had they all accepted the offer. This fear proved to be a major miscalculation. Only 88 of the 570 Thames operatives elected to transfer to Kemsley Mill. When the project had to be put back for two months because of the intervention of the three-day week (a consequence of the miners' strike early in 1974) it was difficult for Bowater to maintain production at Thames without offering special payments to men to stay on. Three things combined to frustrate the company's ready assumption that the bulk of the labour force would agree to be transferred: the attraction of the lump-sum redundancy payments; the buoyancy of the local labour market at the time; and the deterrent of an extra 25-mile journey to work at Kemsley. Indeed, a sidelight on the last factor is that the company had to buy housing for all the operatives who did transfer.

The National Coal Board

25. As a nationalised industry, the National Coal Board is arguably more shielded from the impact of changing market conditions than private enterprise concerns. Its destinies are guided not only by commercial considerations as perceived by its own management, but also by governments sensitive to political influences and therefore willing to subsidise and protect the employment of a large labour force that is also a powerful interest group. As a result, the industry has in recent years been given almost total protection from imported coal (apart from certain specialised grades), governments have on occasion instructed the electricity generating industry—also State-owned—to build coal-fired power stations rather than choosing its own energy sources, and large accumulated debts have been written off. Nevertheless, even governments have to consider the costs to the national economy of relying on a more expensive form of energy, and as oil began to become so much the cheaper alternative in the 1950s, drastic changes did occur in the coal industry. In 1952 NCB deep-mined coal production reached a peak of 210.6 million tons; by 1975 it was down to 112.6 million tons. Over the same period the labour force was run down from 710,000 to under a quarter of a million. Although coal production has stabilised as a result of the sharply increased price of oil and is expected to expand with the opening up of new coalfields, the search for cheaper methods of production has been a managerial priority since at least the early 1960s and this is not likely to end simply because of the upturn in the prospects for the industry. In practice this means the concentration of coal production on low-cost, high-output pits, the closing of uneconomic faces and the development of mechanised mining techniques. This is a large, complex and interactive process, and this study concentrates on how it has been handled over a period in one area, North Staffordshire.

26. Change in this area has been conditioned by the experience of the industry as a whole and the institutions it has developed to handle its problems. There have, of course, been spectacular confrontations in the coal industry, but these have arisen in recent years at national level over pay, while the innumerable local disputes that once plagued the industry

have been largely eliminated. In the coal fields labour relations are now harmonious, thanks largely to the progressive elimination of piecework and the existence of a dominant industrial union, the National Union of Mineworkers.

27. The piecework system was the main cause of the huge number of local disputes suffered by the industry. Face workers were paid only for what they produced and rates were fixed by a process of almost continuous negotiation, pit by pit and face by face. The NUM, which organises all the industry's production workers, has had a policy of substituting a national day wage system for piecework at least since 1946. They achieved this aim in a series of three national agreements in 1955, 1966 and 1971. Of these, perhaps the most important was the 1966 Power Loading Agreement, which ended piecework for the key faceworkers, and thereby cleared the way for the rapid introduction of mechanised mining. The policy of equal rates for equal work throughout the industry was modified in 1978 by the introduction of locally negotiated productivity payments, but it did not lead to a return to strife comparable with that of the piecework era.

28. An increasingly comprehensive system of joint consultation has been established at all levels within the industry since the first top-level National Consultative Council was set up with nationalisation in 1946. With the ending of the piecework system, which occupied so much of the union's local energies, consultative machinery has been further extended to the point where regular meetings are held between management and men at districts within each colliery. Teach-ins and face conferences are held whenever new production problems arise.

29. Although, as with other nationalised industries, strategic policy is decided by a complex political process in which the union, although having a voice, is not decisive, in the day-to-day management of the industry no significant changes are made without the full acceptance of the union and its members.

30. For the purpose of illustration, we choose one example from North Staffordshire, that of the Chatterley Whitfield colliery. In February 1976 it was connected underground with another pit, Wolstanton. This meant that all the coal mined in the two pits was brought to the surface at Wolstanton, although the Chatterley Whitfield men continued to descend from the same pithead as before, and work the same seams. But some mining activity at Chatterley Whitfield ceased as a result of the link-up, and over an eighteen-month period the workforce was run down from 881 men to 385.

31. The first discussions took place at NCB area level, and through them local management and union were aware of what was intended well in advance. At first the NUM branch contested the rundown, but in the end they came to accept its inevitability through mass meetings at the local miners' welfare hall addressed by both union and management representatives. The case for the rundown was explained, and the men were reassured that they would be transferred to other pits or offered redundancy payments as they chose. A week or so later, each man was interviewed during working hours by the colliery manager with the union branch secretary present. The process was not hurried: some men were seen three times before they felt able to make their final decision. If a miner thought he would like to remain in the industry he was taken to visit other pits and offered work at the one that suited him best.

32. The consultative machinery developed in the industry over the last thirty years appears capable of resolving most of the lesser changes that arise, as well as the large-scale run down. Further examples from North Staffordshire illustrate its flexibility. On another occasion at Chatterley Whitfield colliery an automatic electric winder was ordered to replace an old steam winding engine, with the result that 12 jobs on the old steam boiler were lost. At the Hem Heath colliery 36 men were employed as transfer point attendants. Automatic equipment reduced the labour requirement by a third. In both cases the displaced miners were smoothly absorbed without opposition or resistance—even with a certain amount of relief since technology was seen to be sparing men some of the most tedious work. As part of the Coal Board's policy of

keeping work teams together, a face team at the Florence pit is still composed, at their own request, of 28 miners whose own colliery, Foxfield, was closed a decade ago, an indication of the responsiveness of the system.

Debenhams

33. As a department store group, the market judges Debenhams on two criteria: the first is the use it makes of its sites—it has 56 units scattered throughout England, many of them in prime city-centre locations. The second is its efficiency in its business as a large-scale retailer. By the time the company initiated the major changes we shall describe, it had been found wanting by the Stock Exchange on both counts, and its share price plummeted. When the Debenhams crisis arrived, investors' thinking was heavily influenced by a boom in property values. The view was also taken that the stores group had failed to respond to new trends—for instance, the layout of the stores was seen as old-fashioned; departments within the stores appeared to be placed where they were for no better reasons than that they had always been there; in short, that valuable space was being wasted. This analysis was also held to reflect the group's failure to keep up with the spending habits of its traditional middle-class customers, or to attract the custom of working-class shoppers with increased spending power. Middle-class customers, in their search for value for money, were increasingly turning to the chain stores like Marks & Spencer and Littlewoods, which in any case already had the allegiance of the now-prospering working-class clientele. It was in the process of modernising and rationalising its stores in order to satisfy its critics that the company embarked on the most drastic and rapid manpower economies of any of the enterprises in our study. Nearly a third of its labour was shed in fourteen months.

34. In retailing, the labour force is characterised by a high degree of turnover, a large number of part-time employees, and a low level of union involvement. At Debenhams an annual labour turnover of 30 per cent was not uncommon, so that when the company came to attack the problem of costs by staff reductions it could reasonably expect that a large part

of the slimming process would look after itself through natural wastage. In the event, 45 per cent of the labour reduction was achieved in this way.

35. Furthermore, 40 per cent of the staff was part-time, a high proportion of them married women. It was clearly easier to persuade employees with a smaller stake in their jobs to accept fewer hours, re-allocation of time, or even redundancy. Nor did the company have to take into account the objections of a large and powerful union angered by the loss of members: the Union of Shop, Distributive and Allied Workers (USDAW) had less than a tenth of the employees in membership, and it had no negotiating rights, although shop stewards were recognised in a few of the stores. The principal mechanism for consultation was a staff council system established only in 1972, with no power to negotiate even on such a central question as wages. Debenhams' approach to change was magisterial.

36. The company chose a 16 ı per cent cost to turnover target, and then delegated to regional and store directors the task of achieving it. As part of the process, staff levels in each of the stores were critically examined. Although staff cuts varied from store to store, certain principles were followed throughout the company.

37. Employees already over retirement age were asked to go first. Then part-time workers were asked to accept cuts in their hours so as to keep as many jobs open as possible for full-timers. Only when these means of staff saving had been exhausted did the axe fall on full-time employees, beginning with those who were considered to be contributing little to the store.

38. Unlike any of the other case studies, there was no preliminary discussion either with the union or with the staff council representatives at Debenhams, although in one store the director announced the cutback at one of the training sessions at which the sales staff were conveniently assembled. In no instances does it appear that the staff councils, when informed of the company's plans, reacted by suggesting some alternative solution.

39. In each store, each individual was interviewed, and in some cases transfers were offered as an alternative to redundancy. Individual cases of hardship were also sympathetically considered: one girl on the point of being made redundant was retained when it was discovered that she had suffered a nervous breakdown before joining the company.

40. One feature of the Debenhams plan was the remarkable speed with which it was carried out. Within the one-year timescale set by the management, the workforce had been run down by the nearly one-third it had aimed for, with the biggest cuts falling on full-time employees, whose numbers actually went down by 40.5 per cent—from 14,300 to 8,400.

The Financial Times

41. Like any other business, a newspaper can be buffeted by several forms of market force not only in the paper's circulation, the traditional touchstone of newspaper success (indeed, in this respect, the *Financial Times* has been extremely successful and has shown a continuing and steady growth), but in its advertising revenue as well. Far from having a monopoly in its specialist field—business and City news—the *Financial Times* has seen its own success attract other newspapers into elaborate competition, with several pages, and sometimes special sections, being produced to attract readers interested in economic and business news and to bring in the advertising that goes with financial journalism. To meet the competition from other newspapers, the *Financial Times* has been forced to expand its coverage of such areas as general news and the arts. On top of this, its relatively greater dependence on advertising revenue for its income—the ratio is 4:1 in favour of advertising—has made it more vulnerable to general economic recession, which inevitably bites into advertising budgets. And finally, alongside these problems of selling in its own markets—cover price sales and advertising space—the *Financial Times* has suffered as a customer in the market for its principal raw material, newsprint, as a result of spiralling supplier prices exacerbated by a then rapidly depreciating sterling exchange rate. Squeezed by all these forces, at a time when new

printing technology offered the prospect of dramatic labour saving, the *Financial Times* chose this course as the logical escape route.

42. However, the *Financial Times* is part of the British national newspaper industry—often known collectively as "Fleet Street". The industry's products circulate throughout the United Kingdom; its member newspapers belong to the Newspaper Publishers' Association; (although one of the most important, the Mirror Group, had withdrawn) and each newspaper's employees are usually in membership of one or another of seven unions. The newspapers are produced within a small geographical area (a radius drawn half a mile from a point in Fleet Street itself would catch all of them), and thus rates paid and working practices accepted by different newspaper houses are quickly known throughout the industry. The fierce competition that has characterised the newspapers for decades has taken many forms—including labour competition. Managements have conceded claims time and again so as to be sure that their newspapers were published and thus did not lose ground to a rival. This was tolerable as long as newspaper revenues remained high. But it had become increasingly obvious that wasteful labour practices did not provide a sound cost basis on which newspapers could survive through recession. In 1964 a union-management board was set up under a judge, and in 1970 the board was succeeded by a "steering group". But neither, for various reasons, succeeded in resolving the underlying problems of overmanning in the industry.

43. The *Financial Times* thus approached change against an unpromising background. It was a member of an employers' organisation which was itself riven with competitive rivalries; ample experience over the previous decade had shown how resistant to change involving labour reductions the industry was; and the circumstances that were impelling it towards change—falling revenues, rising costs and the appearance of cost-saving techniques—were also bearing upon other managements. It was likely that its efforts to achieve change would become entangled with others, and so it proved.

44. In 1975 a new management team at the *Financial Times* proposed a programme of radical change. Making use of the new computer technology the labour force would be gradually cut by a third. Generous terms would be offered to redundant staff, including the preservation of incomes at existing levels until they found new, alternative work.

45. Perhaps the most radical feature of the programme was the way in which the *Financial Times* proposed to tackle the problem of multi-unionism. In union terms the chief feature of the new technology was that it cut across long-established craft boundaries which coincided with union demarcations. Hot-metal typesetting and the composition of type into columns and pages had hitherto been the exclusive preserve of highly skilled men whose union, the National Graphical Association, had jealously guarded their sole right to do so. Computer typesetting, however, effectively deskilled the job. In future it would be possible for competent journalists or secretaries (both organised in different unions) to prepare type. Yet other unions would be affected by the change, such as those whose members maintained and serviced equipment.

46. Thus the changes would not only affect the jobs and livelihoods of individuals; it also affected the income and the status of their unions. In order to meet this problem the *Financial Times* proposed a new joint technology section—in effect a new union in embryo. All workers affected by the new technology would become members of the new section, within which all pre-existing demarcations would cease to exist. The workers would remain in membership of their old unions, which would continue to receive their subscriptions, and the costs of the new section would be borne by the *Financial Times.*

47. Management took great pains to consult over their plans. They discussed them with the union leadership and also with the leaders of the whole British union movement, the TUC. Great care was taken to explain the plan to the newspaper's staff, through pamphlets and audio-visual presentations. The management's willingness to take such pains was not only a recognition of the probable strength of union reluctance; it also coincided with the fact that it had

time available. For, while the new technology was already in use elsewhere in the printing industry, it was still in the process of evolution and development to meet the needs of the *Financial Times* itself.

48. But although the *Financial Times*' plans appeared to enjoy an encouraging initial response, they were swiftly overtaken by events beyond its individual control. A government-sponsored inquiry into the state of the press in general had investigated the factors that had persuaded the *Financial Times* to undertake radical change, and had concluded that discussions were needed throughout the whole of Fleet Street. A joint standing committee of unions and managements was set up, and in December 1976 it proposed a package of measures to ease the social impact of change, a package not unlike that which the *Financial Times* had proposed for its own staff. Not unnaturally, the *Financial Times*' staff held back from making a deal with management to ensure that they did no worse than others in Fleet Street. In the limbo of uncertainty that then existed the newspaper was hit by outbreaks of interunion disputes.

49. Although national union officials recommended the Fleet Street-wide agreement they had made with the employers, it was in the event thrown out by ballots of union members. Each newspaper in the industry was thus left to make what arrangements it could with its own staff and their local unions. Meanwhile, technical snags in the new technology had somewhat reduced the sense of urgency with which managements now thought of it, while an upturn in advertising revenues had also made its introduction less urgent. By mid-1978, a modified form of the new technology had been introduced at only one newspaper house, the Mirror group (not a party to the industry-wide negotiations). Besides perfecting the technology, multi-unionism remained the outstanding problem to be resolved. Two of the seven unions in Fleet Street looked as if they were heading towards a merger, but in the event it was voted down. The *Financial Times* shelved plans for photocomposition and opted for another form of change—a new European edition.

British Steel Corporation

50. Although, like the National Coal Board, the British Steel Corporation is nationalised and subject to political influence, nevertheless it has not been afforded the same kind of protection enjoyed by the coal industry. BSC sells extensively abroad, and there is a virtually free market for manufacturers to import steel as they need. Following its final nationalisation in 1967 (the industry's post-war history has been that of a political football) a major review of prospects was published in 1973, together with a 10-year strategy of modernisation and rationalisation. If the industry was to have a viable, internationally competitive future, this review concluded, it would have to cut the heavy costs that had grown up as a result of antiquated plant, fragmented production and gross overmanning. Clearly, each of these factors had powerful "social" implications. Apart from anything else, the plan foresaw a rundown of 50,000 jobs throughout BSC, in addition to the loss of 30,000 jobs that had already occurred since nationalisation. Political pressures led to the assignment of Lord Beswick, as Minister of State, to undertake a review of the review, the upshot of which was to reduce the number of jobs to be lost. In the event, what started as a cool appraisal and then became a political issue was rapidly overtaken by the realities of the recession.

51. There were a number of different aspects to the setting up of a Scottish Division within the BSC, involving plant closures, investment in new technology, the shedding and re-allocation of manual workers, and the rationalisation of white-collar staffs. From this range, we first deal with the re-organisation of white-collar workers at a time of intense competition between unions.

52. The statutes setting up nationalised industries in Britain invariably contain a clause requiring the new corporation to recognise and negotiate with trade unions. In meeting this statutory requirement, the British Steel Corporation required its management and white-collar employees to join a trade union. Up until then only a few scientists and technicians among the BSC's Scottish staff belonged to a TUC-affiliated union; most of the rest belonged to a staff

association set up exclusively to represent members in the old prenationalisation private company.

53. Although union membership was encouraged among all employees, no particular union was specified for the white-collar staff. This left a vacuum into which competing unions hungry for members were only too eager to rush. For managers there appeared to be an easy solution. They could, and did, join the Steel Industry Management Association (SIMA) which was set up subsequent to public ownership. For the remaining white-collar employees the choice was not so clear cut.

Within the TUC a special Steel Committee (TUCSIC) had been established by some unions already in the industry, and the BSC would clearly have preferred its non-unionised staff to join one of the TUCSIC unions. One union in this category to offer membership to white-collar staff was the Iron and Steel Trades Confederation, the largest manual workers' union in the industry, which at the time had no great appeal to the staff. The ISTC was itself dilatory in its recruitment efforts, so the way was opened for non-TUCSIC unions to move in. Two unions, the Association of Scientific, Technical and Managerial Staffs (ASTMS) and the Association of Professional, Executive and Clerical Staffs (APEX), quickly did so.

54. In Scotland the outcome of the ensuing struggle for membership left a muddled picture of union representation. As few as 600-700 managers and staff opted for unions belonging to TUCSIC, although the BSC had signed an agreement to negotiate only with TUCSIC unions. The new APEX members and the new ASTMS members were in the extraordinary position of having fulfilled the BSC's requirement that they should join a union, but had chosen unions denied recognition rights by the agreement between TUCSIC and the BSC.

55. In 1969 there developed what has been described as a "God Almighty recognition battle" to resolve which the BSC proposed what it hoped would be a lasting compromise. It agreed to grant recognition to APEX and ASTMS in places where a majority of staff belonged to it and no other union

had previously been granted rights to negotiate on their behalf. This gave APEX and ASTMS opportunities to gain recognition at various establishments in Scotland, and the compromise held up until re-organisation began with the creation of the new Scottish Division.

56. The administrative changes included the closure of the old commercial office of the BSC's General Steels Division and the transfer of some 300 staff, all APEX members, to locations where the ISTC already had negotiating rights for the same grades of staff. A dispute which took the form of non-cooperation broke out almost at once, not, as might be expected, over redundancies among clerical workers—which were as high as 20 per cent—but over the insistence by ISTC members that they should continue to be represented by that union, and a demand by APEX members that their union, too, should be recognised at the offices to which they had been transferred.

57. The BSC found itself powerless to sort out these competing claims—it was still bound by its own earlier compromise solution—so it was left to the TUC to try to resolve the difference between its own member unions. This proved to be a long, and at the time of our study intractable, task.

58. There are other reasons why the pattern of unionisation within the BSC remains confused. SIMA, excluded from the works council and employee director system set up by the BSC, began discussions on a possible merger with the ISTC. However, while such a move would probably please the BSC, it is not welcomed by some other unions. For example, the white-collar section of the Amalgamated Union of Engineering Workers, known by the acronym TASS, does not relish the idea that those who manage and supervise their work should belong to a union with which they might find themselves in dispute, for instance over job demarcation. So TASS would prefer these managers to belong to their own union.

59. It is not only among white-collar staff that interunionism has complicated the processes of change in British Steel. It is also a major factor to be taken into account in dealings

with manual workers, whose representation is spread among thirteen different unions. The major union is the ISTC (which also, as we have pointed out, competes for white-collar membership) with 13,000 process workers in Scotland in membership. The remainder of the blue-collar labour force, maintenance, transport and service staff, are divided among a dozen craft unions which organise themselves into joint committees at national, regional and plant level. These committees, though convenient for welding the members of so many unions into negotiating units, are not always stable, and in Scotland disagreements have sometimes led to one or other of the unions withdrawing from the committees for varying periods of time.

60. BSC's re-equipment programme has been carried out against this union background, and it has also been accompanied by some public debate on manning levels. A well-publicised television programme drew attention to the fact that a plant in Wales (Llanwern) employed 9,000 workers, while an almost identical steelworks in Belgium (Sidmar) was run by only 6,000. These overall figures, however, tend to conceal a major difference which has been a focus of management-union negotiations in Scotland—the use of outside contractors. British unions—particularly the craft unions—fiercely resist the use by BSC of contractors. One obvious reason for this is that the viability of their joint committees as credible negotiating units within the steel plants would be threatened if their functions were to be hived off. So manning levels look higher than in European plants, even though some 500 contractors' men at Sidmar, for example, are available on permanent standby.

61. Nevertheless, a dialogue has developed in Scotland on the correct and acceptable manning levels for new plant, particularly where precise comparisons can be made. For example, at Hunterston, where British Steel planned a direct reduction pelletised iron two-module plant purchased from Korf in Germany, a joint management-union study team visited Hamburg to examine manning at the existing Korf plant. They evaluated such factors as safety, training, the use of contractors and craft demarcation rules (for example, the study team found that foremen in Germany were allowed to

work with tools, while in Scotland they are not). From this detailed comparison the team concluded that, while the manpower requirement at Hunterston might be slightly higher than in Hamburg, the gap would not be nearly as wide as it appeared to be between Llanwern and Sidmar.

62. The tradition of step-by-step negotiation across the board by management and a number of unions on every aspect of the installation of new plant may, in fact, be held to lead inevitably to overmanning. By contrast, however, in meeting the challenge presented by the recessionary rundown of labour, the steel unions found themselves in a quite different role. At plant level the problem was not so much to preserve the maximum number of jobs but to restrain their members from volunteering in excessive numbers for redundancy. Early in 1976 the Scottish division of BSC was faced with a need to reduce its labour force by an additional 1,500 workers because of the deepened recession, of whom 1,400 were in heavy products work. The unions stuck by their national policies of no compulsory redundancies, but agreements with BSC provide payments for those voluntarily leaving the industry at 50 per cent above the levels laid down by the state redundancy scheme. The payments proved irresistible to many workers and created an embarrassing rush to depart. Men not offered redundancy in older plants felt cheated; workers offered a transfer declined to go to new jobs unless first made redundant; in one case, a hundred jobs were ended at one plant and 140 offered at another nearby; yet not one of the displaced workers agreed to move, preferring instead to collect redundancy compensation.

Company Y

63. This company is a manufacturing concern in the north of England. Founded after the war, the firm had enormous financial success in the early 1960s but then ran into severe competition. It had begun the process of surmounting management and labour problems and consolidating its position in the market when it was critically assailed by an event outside its control: the oil crisis had a direct and sudden effect on the sales of its energy-related products. Almost

overnight, production fell by more than half, and management was faced with little alternative but to cut the labour force correspondingly.

64. Events at Company Y fell into three distinct phases. First there were the problems of success, when the company grew so rapidly that its management structure had to be re-organised. Outside consultants were brought in, and they moved on to reform a chaotic wages structure which was souring industrial relations within the firm. In doing so, they laid the foundations for a full system of worker-management consultation. This change was to serve the company well in the second phase, when the market forces described above compelled it to declare substantial redundancies. Cooperation reached a high point when management and shop stewards together lobbied in Whiehall on behalf of their industry. The third phase, however, showed a return to poor labour relations with the shop stewards complaining that, having been willing to consult in times of adversity, the company was no longer interested in doing so when the crisis passed.

65. One aspect of change at Company Y underlines a problem of considerable importance in the management of change. During the second phase, when communications had been greatly improved and real cooperation was developing between the shop floor and the senior management, the company's foremen and line supervisors became alienated and resentful over what they regarded as their exclusion from the new system. In fact, this possibility had been recognised by the consultant and steps taken to integrate the junior management into the system. Unfortunately, the situation was bedevilled by outside events, in particular, the series of national incomes policies, which eroded the pay differentials they previously enjoyed over the men whose work they supervised. Another irritant was the fact that a certain tension existed because of the union affiliations of the junior management. A number of them were members of a white-collar union, but there were complaints from the shop stewards of the two manual unions that several individuals had previously been members and had left without obeying the letter of the rule book.

66. Ill-feeling between the two groups reached a pitch when, in the new spirit of cooperation, the manual workers' shop stewards requested that they should attend the regular meetings of the company's Production Committee, thus turning it into a Joint Production Committee. Junior management stridently objected on the grounds that they sometimes faced rebukes from their seniors at meetings of this committee, and that if accounts of this type of incident were to reach the shop floor their authority would be further eroded.

Harris & Sheldon

67. This is a medium-sized holding company with fourteen subsidiaries, mainly located in the Midlands. Its products cover a wide range, from sporting guns to bathroom units, lifts, vehicle plants, flower bulbs, fuel filters and shop display equipment. In such a diverse setting, clearly no single set of marketing conditions could affect the company's overall performance. But a newly arrived management team saw that performance was considerably less than the market potential, and it devised a strategy to improve it. The process has taken nearly ten years, and its success has not been due to the break-up of acquisitions for their asset value, but to a solid concentration on operational efficiency.

68. One of the major problems identified from the outset was the existence of inefficient piecework systems among the member companies (common to most of the engineering industry at that time). But piecework was recognised to have the virtue of giving the individual worker a degree of autonomy over his pace of work and his level of earnings. So in devising reform, managements aimed to combine an improvement in efficiency with an alternative of corporate incentives based on the total productivity gain, or loss, of the company. In order to secure the maximum support from the labour force (blue- and white-collar), the group buttressed the new pay system with the establishment of joint consultative committees. These have been given a wide scope of activity—for example, they are notified of the company's investment plans and they monitor the bonus scheme—and as

an agency of two-way communication between each of the companies and its workers. Eleven trade unions are fully recognised (though only two or three are normally present at any one plant), and union membership has grown from 40 per cent to 85 per cent. The unions maintain their traditional negotiating rights over wages and working conditions in parallel with the rights to consultation enjoyed by the local committees without the functions clashing—indeed in one or two of the group's companies, the two systems have merged into a single channel of representation.

69. Although this policy has been introduced and encouraged from the centre, individual companies within the group have been encouraged to develop their own detailed schemes for corporate reward systems and consultative committees to fit their own requirements. An example of the success of this approach is provided by one small company in the group engaged in production with a high skill content. New pay structures were introduced for staff and hourly paid workers, and over three years a gain in productivity of 40 per cent was achieved. Overall, Harris & Sheldon has tripled its turnover and seen its pretax profit grow from £860,000 to £3.2 million, while reducing its labour force—mainly through natural wastage—by five per cent.

70. We now pull together the evidence of the general background against which change takes place, as described in the first section of this chapter, and of the case studies. For convenience we have divided our findings under two headings: Employment Policy and the Industrial Relations System.

III

EMPLOYMENT POLICY

71. The case studies showed that very different circumstances led to the acknowledgement that change was necessary among the seven enterprises: market pressures, technical innovations, raw material exhaustion, new patterns of consumption and so forth. In each case the aims of the enterprise included a determined assault on costs, and in particular on labour costs. It is with the handling of the consequent labour problems that this report is most deeply concerned. What perhaps stands out most starkly is the option so frequently preferred by management: the straightforward dismissal of a proportion of its employees.

72. The record is not entirely uniform. The National Coal Board made every effort to redeploy its labour force which, given the arduous nature of mining, is a scarce resource. Faced with a chronic shortage of labour, the Board tried to find new work for men displaced by re-organisation or closure from their old jobs. And Harris & Sheldon achieved most of its labour run-down through natural wastage over an extended period. But all the other five sought absolute, and in some cases almost immediate, reductions in the number of their employees. Two enterprises, it is true, made open-ended commitments to retain their employees beyond the period of change, but these commitments were far from being unqualified. Bowater offered to keep jobs open for all its manual workers (though not its white-collar staff) when it shut one mill and expanded production at another. But it did not actually have enough available work to go round, and it would have been seriously embarrassing if its workers had taken up the offer in full. Similarly, *The Financial Times* put

forward to its employees a package of proposals which included an offer of no compulsory redundancy. But here, too, the enterprise would have been discountenanced if no employee agreed to quit, since a large part of the purpose of the plan to introduce new printing equipment was precisely to make cheaper the production of the newspaper by the employment of fewer staff. The real incentives it hoped would be accepted were quite different: not the prospect of job security, but the attraction of early pensions and incomes guaranteed at existing levels for those who quit.

73. If such management offers as there were to provide job security rang to some extent hollow, it is only fair to point to a matching hollowness in some of the union demands that jobs should at all costs be preserved. At Bowater, only 88 employees out of 570 actually made the transfer from one mill to the other. And at the Scottish Division of British Steel some workers insisted on being dismissed rather than transferred to new jobs at an adjacent site.

74. A number of reasons for the prevalence of the dismissal option suggest themselves. In some cases—Company Y is an extreme example—the crisis which overtook the enterprise and dictated change was so acute that nothing short of immediate surgery on both output and employment would do. In others, management's fear of conflict if it embarked on radical employment change (a question to which we shall return in the next section) may have resulted in action being delayed until a crisis eventually became inevitable. *The Financial Times*, which had for many years shared the problem of chronic overmanning with the rest of the national newspaper industry, is an obvious case in point. But two other influences of a more generally pervasive character are probably more important than these.

75. The first such influence is that of government policy. As we have noted, recent legislation designed to protect the individual worker has concentrated very heavily on the point at which the employee is threatened with dismissal. The most significant of these rights is contained in the Redundancy Payments Act which provides the worker who loses his job for economic reasons with financial compensation related to

his current earnings and his length of service. The original intention of this legislation was to provide some elementary justice for a worker who had given many years to one employer but who might nevertheless find himself dismissed overnight; and at the same time to persuade workers, in the interest of labour mobility, not to resist dismissal when the requirement for their services had genuinely vanished. But this original intention has been gradually subverted over the years. Any value placed by employers on loyalty and long service has been undermined, contributing to an almost automatic recourse to dismissal as part of the process of change. Unions, for their part, use the statutory minimum redundancy payments as a floor on which to build further claims for compensation. As a result, redundancy has become almost an end in itself, something to be sought as one of the very few ways in which an ordinary wage-earner can acquire a substantial lump-sum in the course of a working life, rather than something to be avoided as in the past. It has thus become more and more difficult to persuade employees to accept new work—even with the same factory or office—at a time of labour re-organisation.

76. The second influence which encourages the use of the dismissal option in tackling labour costs lies in the shape of union organisation and membership. As we have already indicated, unions in Britain are divided into a mixture of craft, industrial and general worker units, although none of these descriptive terms is quite precise—craft unions, for instance, being increasingly willing to accept unskilled members while general unions are equally ready to recruit craftsmen. The upshot is that the typical enterprise will have to deal with a number of unions, each representing more or less well-defined interest groups, when it negotiates change. Each union in such a negotiation will seek to protect the special interests of its own members. Characteristically, it will do so by defining areas of work which its members alone may be allowed to handle. A good example of this is *The Financial Times*, where separate union groupings exist at different points in the production process, a large part of their purpose being to ensure that no members of another union are enabled to work at those points: the setting of type and

the making of photographic blocks are just two of many instances where jobs are carefully reserved for members of a specialised union. When a rationalisation of the workforce is necessary it is, under such circumstances, difficult to redeploy and retrain existing staff to match the new job pattern. The only solution is to dismiss the members of one group (and one union) and take on more members of another union. Demarcations of this kind not only affect employment opportunities at times of change; they are built into the system of industrial training that affects workers from the moment they seek employment. If a youth does not get an apprenticeship before he is 20, he is unlikely to be able to get one later, and if he does manage to acquire craft training he may be too late to qualify for the union card that is necessary for him to practise his new trade.

77. Together, the development of the Redundancy Payments Act and the competitive pattern of union organisation would seem to constitute the major barriers to the creation of viable internal enterprise labour markets whose existence would make the dismissal option a far less attractive choice for management. If they were radically modified, greater job security could be offered to employees; wide opportunities to acquire new skills would be made available; and the community would be spared some at least of the costs of change now forced upon it in the shape of redundant workforces.

78. Some community cost element is, of course, an inevitable component of industrial change. Vigorous internal labour markets coupled with high levels of permanent employment reduce job opportunities available to people entering, or returning to, the general labour market.

In the same way the exercise of the dismissal option creates direct and immediate costs to the community in terms of redundancy compensation and, if the worker does not immediately find other employment, unemployment pay and other social security benefits. While the employer makes a contribution to redundancy and social security payments, the direct on-cost to the community is not a calculation that concerns him in the management of change.

State-run services are also available to help redundant workers to find alternative employment (sometimes by setting up temporary employment bureaux at the workplace), supplemented in several of our case studies by management willingness to use business contacts to try to place in new jobs workers they have dismissed. The British Steel Corporation has started a programme for the financing, construction and leasing of factories in areas where it has felt obliged to close steelworks, or proposes to do so, but it is too soon to evaluate the results, which have hardly been aided by the general economic recession. But even here, a cost to the community is involved in government grants and other aids for new industry in development areas.

79. The chief effort, however, appears in nearly every instance to be exerted at or around the point of dismissal. With some exceptions, manpower planning was regarded in the case studies as an emergency device brought into play to take care of the reduction of labour inseparable from the onset of change, rather than as a continuing managerial function open to the usual bargaining process.

IV

INDUSTRIAL RELATIONS

80. The studies undertaken for this report revealed no single approach by enterprises to meeting the social consequences of change. In each case management's attitude was determined by the institutional machinery which existed to represent employees, and in no two cases were these exactly alike. In fact, the studies covered virtually the whole range of machinery likely to be encountered in British industry. At one extreme, Debenhams had no effective trade union representing any significant number of staff; in the middle, the predominant group of enterprises had to negotiate with a number of different trade unions; and at the other end of the range the National Coal Board had to secure the cooperation only of a single established union. We shall return to the two extremes, but meanwhile we examine the experience of the most numerous group.

81. Among this group, the most obvious characteristic was the lack of consistency in the system of union representation. There was no single, uniform or unambiguous point of contact at which management could start discussions. Each time, it had to invent its own tactics. It had to decide, for example, at which level to approach the unions: to develop new forms of negotiating procedure; and even to stand helplessly by, while the unions argued among themselves for representation rights. These dilemmas were seen at their most acute at Bowater, the *Financial Times* and the British Steel Corporation. At Bowater, management drew up an elaborate game plan for redeploying and reducing its labour force which was largely represented by two major unions. Under "Operation Penthouse", as the plan was known, a

meeting was first sought in secret with a national official of one of these unions, and only after later meetings with local officials of both unions was any information disclosed to the shop stewards and the workforce itself. Significantly, management took care to have its plans virtually complete before it made even its first move towards consultation. The secrecy which surrounded the company's preparations suggests strongly that a degree of mistrust existed towards union representatives which made them unwilling to take officials into their confidence at an early stage in the planning. The result was that the unions' role was confined to negotiating redundancy compensation levels and the terms of removal of those workers transferring to the new location. The unions were, therefore, unable to influence the formative stages of change and were presented with a virtual planning fait accompli.

82. At *The Financial Times*, management's aim was to introduce new labour-saving equipment which cut right across longstanding union demarcations, but it was faced with a very tightly organised structure of craft unions commanding every stage of the existing production process. Since it required the cooperation of its employees throughout the period of re-equipment, there was no question of the enterprise presenting its unions with a fait accompli. At an early stage, *The Financial Times* gave the unions a comprehensive insight into its thinking, which included a scheme for solving the interunion problem by putting all the new categories of work into a separate union section, while individual employees maintained their previous union membership identities, thus guaranteeing the unions' subscription incomes. The reservations and misgivings of the unions within *The Financial Times* over the loss of individual craft identity were still unresolved when the argument was transferred to the level of the national newspaper industry as a whole, as the same problems of technological change began to overtake other newspapers.

83. Representation rights were one of the most controversial aspects of change at the Scottish Division of British Steel. Management proposed a rationalisation of headquarters' staffs involving a substantial number of dismissals. Redun-

dancies were accepted with relatively little resistance from employees. The major source of dissension turned out to be not the actual loss of jobs but the respective rights of several unions to represent the surviving employees. An attempt was made by the TUC to establish spheres of influence between its member unions in the industry, but when the major steel union proved slow to take its opportunity to recruit in new areas of white-collar employment, others moved more swiftly. The resulting confusion proved to be beyond the capacity of the TUC to resolve speedily enough to avoid industrial action.

84. From these examples, one characteristic can be easily identified: the unpredictability of union-management relations in a large enterprise where more than one union is recognised, and where there is no permanent joint negotiating machinery which enjoys the confidence of both sides. Uncertain as to where authority lies in the unions' structure, managements improvise quasi-political strategies, delay change until it is forced on them, keep plans from their employees until the last moment, and sometimes, almost in desperation, feel obliged to propose representative institutions for the unions themselves to adopt. But these are not the only mechanisms to be found. From the case studies, two other types emerge: one where the unions are effectively absent; the other where a single powerful union is present.

85. In either type of situation, change appears to be most easily achieved. At Debenhams where there was no significant union presence at the time of change, management had to do no more than inform staff councils, which it had itself set up, before proceeding with drastic cuts in staffing levels. This process achieved the company's objectives, but as a guide to general practice it is of diminishing value, since fewer and fewer large companies have escaped union penetration. Indeed, Debenhams policy has itself changed since the rundown took place towards a more willing recognition of the shopworkers' union, USDAW, as a negotiating partner. In the National Coal Board, by contrast, an equally smooth management of change at the level of the production unit was achieved, but this time through a close relationship between the management and the National Union of Mineworkers.

Here, advance information was disclosed as a matter of normal practice. When grassroots objections were raised, management and union combined to persuade the workers of the need for change, and together they provided individual counselling aimed at minimising any hardship to mine-workers themselves. Change is endemic in an extractive industry like coal and, particularly since the industry's markets began a sharp decline in the mid-1950s, experience has taught both sides the need for established and predictable procedures for carrying the labour force through unavoidable closures, re-organisations and technical innovations.

86. So far, multi-unionism has emerged from this report as a necessarily negative force, hindering or complicating industrial change. But this is by no means invariably the case. Nationally, most unions encourage their local officials and shop stewards enter into joint working arrangements with other unions, often in the form of joint negotiating committees of shop stewards. Among our case studies, Company Y and Harris & Sheldon appear to provide examples of satisfactory interunion cooperation. The preconditions for this in both enterprises was the laying down by management of clear objectives and the creation of mechanisms for consultation and negotiation through patient persuasion over a period of time. The result for Harris & Sheldon was a steady improvement in productivity and profitability, and a reduced number of disputes. The importance of maintaining the momentum of such institutional arrangements was brought out by Company Y, where the system was installed by outside consultants and carried the company through a major crisis involving the loss of half its production and labour force, only to lose the confidence of the shop floor when it was felt that management was neglecting the new machinery for closer consultation.

V

A NEW FRAMEWORK

87. The seven case studies were largely unaffected by the recent wave of new labour legislation, the ascendant philosophy of which is by giving the unions, as of right, a larger element of participation and involvement within the system, they in turn will demonstrate a greater sense of commitment to, and responsibility for, the running of British industry. To this end, the law has made it easier for unions to recruit and organise, to gain recognition from employers, to operate the closed shop, to be consulted in advance about redundancies, to obtain information they need in formulating claims against employers, to establish themselves in important areas as the single channel of representation for workers, to secure paid time off from work for their lay representatives, and—subject to the passage of new legislation—to sit on the boards of directors of industrial and commercial enterprises.

88. The foundation of a new structure of rules and institutions within which the industrial relations system will operate from now on has thus been created. We cannot judge how this system will work, how, for instance, it might have affected the changes described in the studies had it been installed at the time the changes were made. Would it have made change easier and more readily acceptable, or less so? Procedures would certainly have been different. Management would not have been able to keep their plans as secret for so long as they did in several of our studies. But would the unions use the advance information to which they are now entitled to resist change or facilitate it to the best advantage of their members, the enterprise workforce and the economy as

a whole? Only the experience of a number of years will give the answer one way or another.

89. At this stage we simply note that the unions have themselves been left largely untouched by reform. (The same could be said of employers' organisations, but as we have found, they play at most a marginal role in the management of change.) The unions, as the key component in the industrial relations system, have only been compelled to alter their behaviour in such relatively minor matters as the new requirement that they must open their memberships to either sex instead of, as sometimes in the past, to one sex exclusively. At the same time, the unions have been given a set of major new opportunities to organise, recruit, win recognition from employers and represent their members at all levels in the enterprise and in the economy at large. The test of the new system will, therefore, be twofold: how far employers will be able to accept and adapt to the intended new partnership with their employees, represented through the single channel of the trade unions; and how far the unions, designed above all for an adversary role in collective bargaining, will be able to exploit their opportunities in effective, constructive and responsible ways.

CHANGE IN WEST GERMANY

I

THE BACKGROUND

1. The economic development of the Federal Republic has been characterised by a rapid and, for the most part, steady growth. In addition to strictly economic factors, such as the development of export markets, this growth record can be explained by certain social and political conditions that have prevailed in post-war Germany:

- — the performance of the labour market in meeting constantly expanding demand, firstly, by the influx of mainly skilled workers from the East, and subsequently, by the recruitment of foreign labour—the so-called "guest" workers;
- — policies of taxation and subsidies which encouraged in many sectors of industry an "aggressive" programme of investment.

As well as the favourable investment climate created by such conditions, growth was assisted by the remarkable stability of industrial relations, both at industry level and at the level of the individual enterprise. It is this industrial relations dimension that has increasingly attracted the attention of foreign observers over the past years. The German "economic miracle" appeared to many to be a miracle of industrial relations and, as such, was viewed with a mixture of unease, scepticism and admiration.

2. Initially, attention was drawn to the exceedingly low strike record in the Federal Republic. The German strike rate was consistently lower than in other industrial nations, and this fact was often attributed to the mentality of the German worker. In fact, in our view, it had more to do with

policies adopted by trade unions in the Federal Republic in their negotiations with the representatives of capital.

In its large-scale survey entitled "Trade Unions in the Federal Republic", a study group, set up by the Institute of Social Research, Frankfurt, compares trade unions that adopt an "adversary" approach with unions which follow a cooperative policy. "Cooperative trade unions try to protect the interests of their members by tailoring both wage demands and strategy to the state of the economy and the needs of growth. They take account of the marginal economic concessions available to the other side when formulating their wages policies, and, given the existence of institutionalised incomes policy, agree to abide by the targets and guidelines set by government economic policy. 'Adversary' trade unions, on the other hand, attempt to realise the articulated interests of their members through immediate wage demands and strategies. They refuse to relate their wage demands according to the state of the national economy. Their demands take account of existing market conditions and of their own current strength. They refuse to take responsibility for the national economy as a whole."[1]

3. There can be no doubt that since the mid-1950s the majority of unions in the Federal Republic have predominantly followed a policy of cooperation. This applies particularly to technical and organisational innovation. Rationalisation was regarded by the unions as a necessary and unavoidable part of economic development. "Rationalisation, including mechanisation and automation, create the necessary preconditions for growth and contribute to protecting and improving the standard of living that has been secured."[2]

As a result, the trade unions did not try to interfere with or impede the process of technical change. Rather, they were concerned to see that adverse effects on the workers were compensated for by financial or other means.

The policies which were pursued by the unions in the two decades after 1950 provided a powerful motor for economic expansion. This was so not only because there was little opposition to technical and organisational changes, but also because such changes tended to go hand-in-hand with increased financial benefits for the workers involved. The greater purchasing power that was thereby created acted as a stimulus to increased growth.

4. The "cooperative" policy pursued by the unions was however linked to a number of preconditions, a fact that is often overlooked:

(a) There had been full employment since the mid-1950s. As a result, workers who had been made redundant found new jobs without much difficulty. Indeed, few workers found their way onto the labour market; most could be re-employed within the enterprise. In turn, economic expansion guaranteed full employment.

(b) The way the trade unions were organised, on the principle of one union per industry, ensured that competition between individual unions was eliminated.

(c) The legal-institutional system of industrial relations regulations in the Federal Republic. Two special features characterise the system of industrial relations in the Federal Republic:

    (i) Although the legal and administrative framework guarantees the autonomy of collective bargaining, it both structures and limits the forms and negotiating areas of capital and labour by establishing a multiplicity of rules, which create conditions for controlled management of conflict. The legal framework determines the principles by which negotiations, conciliation and strike action are conducted. The conflict of interests is limited by the high degree of legal codification. Potential clashes are neutralised and the political dangers that are inherent in them are minimised.

    (ii) The Works Council Law ("Betriebsverfassungsgesetz") establishes the works council as an autonomous body representing the workforce within the enterprise, independent of the trade unions.[3] It thereby creates the "dual representation of interests system" which is a characteristic feature of industrial relations in the Federal Republic. The regulation which stipulates that the ordering of internal plant conflict is the affair of the works council contributes to limiting trade union autonomy in collective bargaining.[4]

Additional provisions for labour-directors, nominated by the unions, and parity representation in the supervisory board, were made for the coal and steel industry by the "Co-determination Law" ("Mitbestimmungsgesetz") of 1951.

5. Here we have touched on a point vital to the understanding of industrial relations in the Federal Republic, i.e. the well-defined "division of labour" between local (i.e. enterprise or plant) and industry-wide representation of employee interests.

While the trade unions lay down a binding framework of terms and conditions of employment in their collective agreements, the framework is filled out by negotiation at enterprise or plant level, and thus depends to a considerable extent on the activities of the works council. In other words, in trying to understand industrial relations policy, and the consequences for industrial change, we have to look not only at the activities of the unions, but also at what is happening within the enterprise.

6. The rights and duties of works councils were laid down by law in 1952. In 1972, a new extension of this law came into force containing an extension of the consultation and co-determination rights of works councils.[5] In the planning and implementation of technical-organisational change and the consequent manpower development, the new Works Council Law introduced a number of additional regulations. The most important of these were:

— Co-determination rights given to the works council in cases where the introduction and use of new techniques is designed to control the behaviour and performance of workers. (87)

— Co-determination and consultation rights over the physical environment, the organisation of an individual's work, and working conditions. (90 ff)

— It is the duty of management to inform the works council in good time and in detail of manpower planning, particularly in so far as it affects current and future personnel needs, to enable appropriate measures concerning personnel and retraining to be taken. Management must discuss the time-scale and extent of the measures it intends to adopt with the works council. The avoidance of hardship must also be discussed. (92)

— Management is obliged to inform the works council in cases of manpower restructuring and transfers. In these matters, works council agreement is necessary. (99)

— Management must discuss proposed dismissals with the works council and must give reasons for them. If men are dismissed without its agreement, the action is illegal, and the works council can refuse to accept it on the grounds detailed in the Law. (102)

— Enterprises with more than 100 employees are required to set up an Economic Committee whose membership is determined by the works council. Management must inform this committee about the economic state of the enterprise, and forthcoming plans, including investment, rationalisation, closures, etc. (106)

— Management must provide early and detailed information on any planned changes within the enterprise which may have a negative consequence for the work-force, or for considerable parts of the work-force. It must also discuss such changes with the works council. (111)

7. Besides the provisions of the Works Council Law, the functions of works councils concerning change in the enterprise have been defined in a series of collective agreements between trade unions and Employers' Associations. For example, the agreement concluded in 1968 between I. G. Metall and the Joint Employers' Council which deals with protective measures to be taken in cases of rationalisation stipulates that management must inform the works council when planning rationalisation. This must be done as soon as management knows the extent of transfers, retraining and the

redundancies which are entailed. Any possible social consequences have to be discussed with the works council.

In addition, the agreement provides for the following:

— wherever possible, the employees affected by change should be offered jobs of comparable status to the one they had lost;

— compensation should be paid in cases where older workers with a long service record could not be offered jobs of equal status;

— the payment of financial compensation to workers dismissed as a result of rationalisation; such payment to be graded according to length of service and age.

8. Although such agreements do not provide for co-determination rights, but merely the right to be consulted, taken together with the legal provisions of the new Works Council Law, they represent a considerable strengthening of the position of the works council in the personnel aspects of industrial change.

9. To sum up, during the past twenty years, economic growth, stable conditions of employment, the process of technological and organisational change, and industrial relations have all been closely interrelated in the Federal Republic.

10. The focal point of this system of interdependence is the enterprise. It is the place where rationalisation is carried through, where employment policy[6] is formulated and the effects of rationalisation are handled, and where industrial relations policies are turned into reality.

The enterprise also is the focal point of our investigation, which attempts to clarify the interrelationship between the implementation of technical-organisational change, the employment policies that are followed, and industrial relations as practised by management and employees' representatives.

11. Our study of this three-cornered relationship between manpower policy, innovation and industrial relations appears particularly well-timed in view of the changed economic

climate that has come about as a result of the 1974 economic crisis, and as a consequence of new laws and collective agreements. They confront the works council with new tasks, new rights, and new duties when rationalisation measures are carried through, at a time when the increasingly difficult employment situation has put greater strains on industrial relations.

References

1. J. Bergman, O. Jacobi, W. Müller-Jentsch: "Gewerkschaften in der Bundesrepublik", Frankfurt am Main, 1975. S. 28.
2. Cf. "Abkommen zum Schutz der Arbeiter und Angestellten vor negativen Folgen der Rationalisierung", 19. Mai, 1966, between I. G. Metall and the Employers' Association of the metal industry.
3. It should be noted that works councils are not trade union bodies, though in practice a large proportion of works council members belong to a union. But they are elected by all the employees of an enterprise, including non-unionists.
4. Bergman et al. p. 175.
5. It is necessary to distinguish between rights of consultation on the one hand, and co-determination rights on the other. The former compels the employer to inform or consult the works council, while a measure taken by the enterprise is, with co-determination, only legally valid if the agreement of the works council is forthcoming.
6. By employment policy, we mean the totality of measures taken by the enterprise, and the various regulations governing activities at work. It comprises orthodox personnel policy in a narrow sense, as well as wages policy in so far as this serves the purpose of motivation and influences the behaviour of the workforce.

II

THE CASE STUDIES

12. The problems and enterprises involved in the case studies were:

> conversion of the replacement engine shop at a car factory;
>
> the construction of a universal mill train at an iron and steel works;
>
> the introduction of a different cashing system in a department store;
>
> the change to contract handling without files in an insurance company;
>
> the building of a new stripping plant in a paper factory;
>
> the construction of a new continuous casting line and closure of a slabbing line at an iron and steel works;
>
> the introduction of computerised typesetting in the printing industry.
>
> (for reasons of confidentiality we have been asked not to name any of the enterprises concerned in the studies).

The Car Factory

13. Factory D belongs to a multi-national car production concern. It is located in a medium-sized town with a fundamentally rural catchment area. Being by far the largest employer in the entire region it has a relatively good recruiting situation. This is indicated by the small number of foreign employees (for a car production plant), which has never exceeded 25 per cent of the entire workforce and had dropped to 17 per cent by the end of 1975.

14. In the course of the last 20 years, the company has gone through a number of changes, and production and employment have fluctuated considerably. Following a period of strong expansion in the 60s, the company was very severely hit by the oil crisis. The workforce fell from 19,500 in December 1973 to 15,500 in December 1975, more than 20 per cent.

15. The need for a manpower reduction was foreseen towards the end of 1973. The first steps taken were quite conventional: a ban on new employment, with new hirings confined to individuals who could not be replaced by internal recruitment.

16. It was soon realised, however, that this would not be sufficient. As the primary objective was to avoid mass dismissals, it was decided to give employees a motivation to leave by offering them termination contracts coupled with a settlement payment. Two schemes were organised. The first went from September to the end of November 1974 and led to the conclusion of some 1,300 settlement contracts, the second from June 1975 to June 1976, with 1,700 settlement contracts.

17. The basic manpower problem was to keep this reduction of the workforce under control, and it was hoped simultaneously to improve the age and qualification structure of the workforce by reducing the numbers of older and less qualified employees. The first objective was reached by offering an early retirement programme, which about 800 wage-earners and salaried personnel opted for.

18. Improving the workforce qualification structure proved to be more complicated. An attempt at grading the settlement contracts according to current wages and duration of employment turned out to be unsuitable. It showed that the people most inclined to leave the plant were precisely the ones the company wanted to keep—young, qualified German employees. To avoid this, settlement contracts were made conditional on the applicant first obtaining the agreement of his responsible superior.

19. The works council participated in the settlement and re-organisation scheme from the beginning. With their agreement in principle that the scheme was necessary, the company was able to take the requisite measures quickly and without delay. This was important because a delaying strategy by the works council might have jeopardised the entire cost reduction objective.

20. In fact, the pattern of works council cooperation faltered only at one stage, when management decided to close down the replacement engine shop, where engines for the models manufactured at plant D were repaired and overhauled. In spring 1975, the total workforce in the replacement engine shop was only 60, including 4 foreigners and 13 women. It was decided that the shop should be closed and the work transferred to another plant where more modern equipment was available.

21. Following preliminary talks the works council immediately filed a formal complaint against the proposed closure on the grounds that it breached an agreement on the company's employment structure.

22. As a result, the plant planning department considered various alternative solutions with the objective of maintaining the jobs involved. The result was a proposal to transfer the engine overhaul work as planned, but to use the replacement shop for overhauling electrical components for all plant D models. This proposal was submitted to the joint committee of the concern in May 1975. The proposal was supported by feasibility calculations prepared by the finance department.

23. At first sight it might appear surprising that the works council, which had accepted a workforce reduction of several thousand employees, was not willing to accept a decision which would only have affected a relatively small number of jobs—just as it may seem surprising that management was willing to seek alternative solutions. This becomes understandable, however, in the light of the particular relevance of the replacement components shop and the jobs involved to the overall employment policy of plant D.

24. The replacement components shop differed fundamentally from the central production departments, such as the assembly lines or the mechanical production shop. Owing to the small quantities handled and the multitude of different jobs to be done, it was not worthwhile to introduce a line system. The work was performed at individual work-bays, basically without the time pressure usually generated by a piecework system. As only light jobs had to be performed at a number of work-bays, the employees were able to work sitting down, and there was no shift-work.

25. The shop was not subject to the particular stress otherwise typical of car production lines and so it had a specific function in the plant's employment policy in that it served as a "retreat" for production workers unable to work on the mechanical production lines and not really qualified for transfer to other departments.

26. One consequence of the proposed closure, therefore, was that it threatened to eliminate job opportunities for older and less efficient employees. The availability of jobs for limited-capacity employees is of particular significance at plant D, since a plant agreement concluded in July 1974 had stipulated employment and wages guarantees for old or long-service employees over and above the legal requirements. Closure would thus have impaired the flexibility otherwise required within the plant's employment policy. In the light of this situation, the intervention by the works council did not only represent the interests of the plant's workforce, but was also in line with the company's accepted employment policy.

27. The solution finally found was typical of the current business situation at the time: to increase plant flexibility by cancelling orders given to third companies. As long as there was full employment, such orders had been farmed out to increase the plant's own production volume. Now, with a drop in the level of employment, this work was called back, even though this would not provide a reduction of costs in commercial terms.

28. The replacement components shop case also illustrates the impact of the existing legal regulations. Although, under

current German law, there would not have been any absolute barrier to closing down the shop, the implementation of such a decision would have caused practical difficulties because the structure of the workforce was predetermined by an enforceable legal agreement. The "solution" adopted is thus exemplary not only of the impact such agreements may have, but also of the typical way a company can be persuaded to respond.

Iron and Steel Works (1)

29. Plant A belongs to a steel company which operates two steel mills and ore mines. In 1970, the plant, which up to that time had been held by private shareholders, was integrated in a publicly owned steel concern. Plant A is located in a medium-sized town with a basically rural catchment area. It is by far the largest local employer.

30. The plant concentrates on the production of sectional steel. Its overall steel output, which was 5.5 m tons in 1974, has fluctuated considerably in recent years following the constant change of the steel market. In the 1974/75 business year, steel production went down by 22 per cent as compared with 1973/74, and in the middle of 1975 production was one-third lower than in the same period of the preceding year.

Despite these substantial fluctuations, the workforce remained basically unchanged. The total number of employees, which had been almost 20,000 in mid-1974, only dropped slightly to 18,600 by spring 1976. At the same time, the workforce of plant A decreased from 7,300 to 6,850 in April 1976. This workforce reduction was achieved by imposing a ban on new employment. No employees were dismissed due to shortage of work. The normal working time/overtime ratio was also reduced from 9.5 to 0.6 per cent. The concern thus succeeded in reducing the total number of working hours by the equivalent of almost 2,500 employees, i.e. 12.5 per cent of the active workforce. Short-time working was also introduced for several periods throughout 1975.

31. As the company is subject to the German Co-determination Act the workforce is represented on a pro rata basis

not only in the company's supervisory board, but also in the overriding holding company and in both boards of management (through a labour director). The relatively strong position held by the labour director is underlined by the fact that he is also the deputy spokesman of the 6-member board of management.

Another special position, which exceeds the legal regulations set forth in the Coal and Steel Act, is that of a "labour economics works manager", who cooperates with the technical plant manager in both of the concern's works and is appointed by the labour director in agreement with the trade unions and works councillors. The labour economics works manager is responsible for all of the divisions and departments within the scope of the labour director's activities, unless such divisions or departments perform specific functions for the overall concern.

32. The works council of plant A obviously plays an important and strong role within the plant's labour scene. The chairman of the works council has been in office for five years, his deputy for four years. The plant's workforce is represented by a large number of councils and boards including the overall works council of the holding company; the overall works council of the steel company; the executive works committee of the steel company; the economic committee of the steel company; the works council of plant A; and the works council committee of plant A.

33. In mid-1975 the new universal mill train started production at plant A. Its purpose was to replace the two old rolling trains built towards the end of the 19th century. Although the old rolling trains had constantly been improved over the decades, they did not comply with requirements for several reasons. The two old trains had a maximum monthly output of 11,000 tons in 15 shifts per week, while the new train produces about the same volume in only five weekly shifts and can reach a maximum of 35,000 tons when operating in 15 shifts per week. When expanded to phase II, the new train will even reach a production volume of 60,000 tons in 15 weekly shifts. Its maximum ceiling, when fully expanded, is 85,000 tons.

34. The first of the three old-train shifts was discontinued early in 1975, followed by the second shift in August 1975. Trial operation of the new train started in July and production was officially taken up in November 1975. The changeover to two-shift operation was scheduled for June 1976, by which time production on old trains was to be discontinued completely. A definite date had not been arranged for the start-up of the third shift at the time of this study.

35. As of the beginning of 1975, the total workforce for the old rolling train was 300 employees (250 for production and 50 for maintenance). 250 were wage-earners and 50 were on a salary pay-roll. Out of the 250 wage-earners, 167 were selected for operating the new train, 55 were to be shifted to other departments in the rolling mill, 8 foreigners were to leave the company on account of their work permits expiring, and 26 workers who were all older than 59 were going to leave the company on their own accord and retire ahead of time.

36. When production started, 122 wage-earners and 37 salaried personnel worked on the new train (59 for production and 63 for maintenance). With full three-shift operation the workforce will be increased to 325 people (167 for production and 158 for maintenance). 64 of the original workers came from the dissolved first shift of the old rolling trains, and 48 from other departments. No new wage-earners were hired for the operation of the universal mill train (only four operation managers—who were all put on the salary pay-roll—were added as new employees from outside the company).

37. The core team for the new universal mill train, a total of 84 employees, was prepared for the new job by internal training courses. On an average, each team member received 80 hours of such further training. 27 employees participated in external further training (e.g. with the manufacturers) on a total of 208 days. Although a relatively large number of employees was taken over from the old trains, the change in workforce structure was nevertheless quite considerable. The percentage of foreign workers, which had been about 30 per

cent on the old trains in early 1975, dropped to 12 per cent on the new train. The group of over 53-year-olds, which had made up more than 16 per cent of the old trains' workforce in mid-1974, went down to 4 per cent.

38. While the overall workforce went up, and not down, certain problems did arise in the process, particularly as some of the old train employees were found to be unsuitably qualified for the new train. The old train had required a high level of physical stamina and certain manual skills, whereas workers on the new train had to have other qualifications.

39. To find out how many of the old-train employees were suited for working on the new train, the operations department compiled a list for each individual employee showing his most important personal data. This list was put together by plant management and the responsible supervisors in order to obtain a basic guideline as to who could be used on the new train and who could not. The result was then discussed with a representative of the labour economics department and the works council, with whom a common opinion was established in most cases. In this process the works council succeeded in getting about 10 per cent of the workers originally rejected by plant management on to the new train, after all. Whereas plant management generally gave preference to recruiting new people from outside, i.e. either highly qualified specialists or younger employees, the works council endeavoured to get the old group on to the new mill train.

40. In the presence of the works council, the employees found suitable for the new train were asked by plant management and a representative of the labour economics department where they would like to continue working. Of the 180 employees asked this question, about 30 did not want to switch over to the universal mill train and were offered other jobs in various departments which, however, they did not have to accept. Instead, they also had the possibility of choosing a different job or a different department.

41. One problem was the maintenance of wage levels. It was not possible to make all the changes necessary without reducing the employees' earnings in individual cases.

Although such employees receive their old wages under the basic wages and salary agreement and a specific wage contract concluded by the plant, and although there have not been any wage reductions so far, all parties realise that the current situation is only a transitional one and that some workers may lose up to 10-15 per cent of their original wages.

42. As one objective in introducing the new mill train was to save labour costs, the number of employees, though higher than before, is very low indeed. Management has even accepted the risks this presents in terms of lost production, illness, holidays, downtimes, etc., while at the same time trying to introduce more manpower flexibility by grouping together larger sectors. For example, the employees responsible for conversions and plant modifications now also perform production functions.

43. The works council was involved in the planning process from a relatively early stage; not only the labour director, but also the chairman of the works council participated in the investment discussions held by the board of management. Written information on the envisaged project was then given to the works council, which was asked for its approval, before the matter was referred to the supervisory board. Finally, the ensuing discussion and decision-finding process in the supervisory board was also carried by the employees' representatives, among them the chairman of the works council of plant A.

44. The way the project was described by the parties interviewed, the whole process may be regarded as a model case of successful co-determination which accelerated and improved the project instead of slowing down. The works council, by intervening at the right time, helped avoid planning errors which would otherwise have cost a lot of money later on. This not only involved so-called "social" facilities, but also technical or organisational planning criteria.

45. Though almost all the employees were trade union members, the trade union itself seems to have had only a marginal role within the company. Although trade union

representatives take part in certain meetings, they apparently have a relatively minor influence on what happens at the company. In recent years the parties have been very successful in avoiding the outbreak of open conflicts. Since 1945, only one petition has been lodged with the arbitration board—this was in the 50s and concerned a ban on alcohol management wanted to impose without consulting the works council.

The Department Store

46. Department Store G, which is in the centre of a large city, belongs to one of the big department store groups in the Federal Republic. Both in the group and in Department Store G the last few decades have seen no strikes or other open conflicts between the staff and management. Cooperation between the management and the works council was also relatively free from conflict. An important factor may have been the high degree of continuity among the holders of the top offices both in the works council for the concern as a whole and in that for Store G. The predecessor of the present chairman of the works council in the store held the office for 17 years. Like his successor he was a departmental manager. (He was not re-elected in 1972. The reason was said to be dissatisfaction among the staff at his policy, which was too favourable to the company. His successor is said to be more active.) The degree of union organisation among the staff, two-thirds of whom are women, is relatively low.

47. The change in the cash system in Store G was an attempt to achieve a redistribution of functions and above all a reduction in staff through a re-organisation of the sales, cash desks and packing.

However, different ways of achieving this aim were used in different departments. For example, on the fashion sales floor, the procedure was that the sales assistant was there only to advise the customer and make the sale. The article was then taken by the sales assistant or a junior to a central packing table and cash desk, where different staff took the money and packed the article. These central cash desks and packing tables served departments so that they were not under the supervision of one department but the central administration. Now that new cash desks have been installed the sales assistant, after making the sale, takes the article to the cash desk, where she takes the money and packs the goods. Additional staff are only put on the cash desks when the store is very busy. The cash desks are as close as possible to the sales area and are under the supervisors of the various departments.

48. In the cosmetics department, as in others handling small articles which do not need much packing, the goods were sold, packed and the money taken by the sales assistant. There were a large number of cash registers available in the sales area.

49. The change provided one central cash desk for a large part of the sales area and the removal of the many smaller cash registers. The customer takes the article without a ticket from the sales assistant to the central cash desk, pays there and has his article packed. The aim of this innovation was not only to reduce the staff required but also to simplify the sales process and eliminate the mistakes in the many decentralised cash registers.

50. Altogether the store expected to save between 80 and 100 jobs through the change in the cash system in the two departments. Up to the time of this study, the spring of 1976, however, only seven jobs had been saved through re-organisation of the cash system.

51. In a discussion with the management of the store, the personnel manager and the organisation manager, it was decided to make the first changeover in the fashion department. After this discussion the works council was officially informed for the first time. A copy of the rough plan of the measures was given.

However, the works council, at least its chairman, had already been informed of the basic outline of the measures at one of his regular meetings with the management. He was told that the matter was not yet ripe for discussion and the management promised further information shortly. Further news soon followed in the form of personal preliminary information from the personnel manager, and finally a regional meeting was held for the works council as a whole, at which reports were given of similar measures in other stores. It was emphasised that there were to be no redundancies. After a detailed discussion the works Committee approved the measure on this Condition.

52. The works council only regarded action on its part as necessary if there were to be transfers, regrading or redundancies. It did not believe that it should play a part in the actual organisational planning process, but only to make sure that certain minimum requirements regarding job security and maintainance of earnings levels were fulfilled.

53. At first, redundancies or transfers were avoided because departments with a "favourable employment structure" were chosen—those in which cashiers or packers were already due to retire. Secondly, a "salami technique" was deliberately used for the changeover, with the re-organisation carried out step by step, one area after another. This meant that the whole process took a relatively long time. A great advantage was the reduction of conflict and tension. In many cases there was no need for decision by the works council as staff changes (transfers, regrading or redundancies) were not needed.

54. Until 1973 the policy of the group had been clearly expansionary. The main aim was to acquire a higher market share through extending the sales area and increasing the number employed. The main problems in employment policy at that time were recruitment to achieve the number required, training for the new staff and reducing the fluctuation rate. The tempo of expansion was limited by the finance available and also by the situation on the labour market. It became increasingly difficult to find enough suitable staff. The fact that largely women, many of them part-time, were employed meant that in this sector recruitment was possible longer than in other sectors of the economy, but in the period of full employment, from 1968 to 1973, even this potential began to fail. The use of foreign labour, which had provided a solution in other sectors, was possible only to a limited extent and in the sales area not at all.

55. All these factors made it necessary to revise the policy of expansion at any price in favour of a more intensive sales

policy, i.e. the achievement of higher turnover and profits on the same floor space and with fewer staff. The oil crisis at the end of 1973 gave fresh impetus to this concept. The stagnation in sales volume with rising costs made it more necessary than ever to introduce cost-saving measures. The reduction in sales staff was bound to be the focal point as this accounted for two-thirds of costs.

56. Those concerned, the personnel management, the planning department, the works council, did not question the need for the rationalisation measures. Doubts were only expressed as to whether the necessary gain in productivity could be achieved. No alternative possibilities, which might be better from staff point of view, were examined. The framework or out-line plan worked out by the central planning department was never questioned. It has always been accepted as "the" solution.

57. For the works council the avoidance of redundancies and regrading in a lower wage group were the main criteria. This did not apply, however, to a reduction in the number of part-time staff which did not involve any labour legislation or the approval of the works council as there were no contractual arrangements with such staff. Store G had employed large numbers of part-time and occasional staff over the last few years, so that this seemed a promising solution. The store had 2,200 full-time staff and 383 part-time. These were reduced to 109.

58. One other staff reduction technique was seriously considered by management: an unobstrusive "purge" of "undesirable elements" among the staff. The works council and the personnel management seemed largely agreed on the definition of "undesirable".

59. In fact the works council has always accepted a disciplinary function. It plays a part in the bag check when the staff are leaving the store, it is responsible for this being properly carried out and it has to defend the measure to the staff at general meetings.

60. In view of this attitude and behaviour on the part of the works council, it is hardly surprising that both the

management and the works council emphasised that intervention by the latter has not had any inhibiting effect on the rationalisation measures.

61. If this case seems to show the limits to the influence of the works council, it must be recognised that this high degree of cooperation is particularly important in a department store where there is always the risk that employment policy will seep through to relations with the customers. Even more than any other business, Department Store G does not want any tarnishing of its public image.

The Insurance Company

62. Company F is one of the largest insurance concerns in the Federal Republic. The head office employs about 1,300. 10,200 are employed in the various offices and administration centres in which 14 million insurance contracts are actually handled. A further 1,300 are employed in the head offices and offices which plan the deployment of the 52,000 members of the external service.

63. During the fifties and sixties the insurance group was expanding steadily. The number employed rose from 6,300 in 1950 to 18,600 in 1971. At this time, three years before the oil crisis, however, the number employed had already reached its peak. By the beginning of 1976 the number had dropped steadily to 17,140. The reduction was achieved mainly through a net stop on appointments since 1971: only staff who left could, if necessary, be replaced, but no additional staff could be taken on. No one was declared redundant through rationalisation or lack of work.

64. The administrative central office, in which the experiment of contract handling without files was mainly carried out, is in a large town in southern Germany. It resulted from an "administration reform" in the concern, when the branches of several subsidiaries were combined in 1969. It employs about 2,000, half of them women.

65. The works council in the branch has 19 members (17 white-collar workers and two blue-collar workers). The chairman has been in office for 15 years and has been a

member of the works council for 26 years. He is at the same time chairman of the works council for the subsidiary as a whole. About 10 per cent of the staff in the branch are in unions. The use of data sorting equipment and the gradual transition to contract handling without files was the last step in a long process of continuous change in contract handling in the Special Department for Retail Business in branch M.

66. The rationalisation process can be subdivided into three periods:

(a) In the first, which lasted approximately from 1959 to 1964, the emphasis was on "conventional" rationalisation, i.e. changes were introduced in the forms to make the work process smoother, the registry more compact and easier to use. But in each case a file had to be made up, then used for all subsequent handling.
(b) In the second period, which lasted from about 1964 to 1973, the essential information was transferred from the files onto punch cards, which were then used for subsequent work. The registry was replaced by punch cards and the microfilming of other information, which saved a considerable amount of space.
(c) It was only a small step theoretically to move from this system to the use of a computer to retrieve the information contained on the punch cards. With the development of data sorting equipment at the beginning of the seventies the technical prerequisites were also available. This was the third phase of the rationalisation process. It was a gradual process, in which more and more work processes were gradually transferred. The most important prerequisite was the development of efficient screens, which came onto the market in 1974. Only then was it possible to do without files and punch cards and to have the contracts handled entirely in a dialogue with the computer.

67. The work is organised in groups. A group of six clerks each has a certain number of contracts to look after. On principle all the work involved is done in that group, so there is no division of labour between the groups. Within the group there is a certain amount of specialisation, but the aim is to

keep the work as broad as possible. Each group has two screens. Work on the screen is arranged by agreement within the group; each clerk spends about a quarter to a third of his time on the screen. No further fundamental changes, except for "polishing or rounding-off", are planned to this process. At present about 80 per cent of the work is done by dialogue with the computer and the aim is 100 per cent.

68. The introduction of data sorting equipment in the Special Department for Retail Business and the transition to contract handling without files was only one—even if a very advanced part—of a rationalisation process which brought a drastic increase in productivity: the number of policies handled by each member of the staff rose in this area from 13,000 to 22,000.

69. For the Special Department this meant a drastic reduction in staff. In 1970, 110 were employed in the department, and in 1972, before the data sorting equipment was introduced, there were 97. By 1976 there were only 45. This reduction was achieved without redundancies. The early date of the appointments stop was certainly an important factor: it not only brought a very much longer period of "natural wastage", but also meant that the reduction progressed much more rapidly in the years of full employment up to 1973-4 through voluntary departures.

70. What is striking is the "quietness" with which the process of technical and organisational innovation and rationalisation was carried out. Most of the individual organisational changes seem to have been brought about without formalised approval procedures—for instance in the form of detailed presentations to the board. The basic decision on such projects appears to have been made by the Commercial Planning Department and the Board member competent for data processing.

71. This is partly due to the fact that the investment sums involved are relatively low in comparison with those for wage costs. But the "quietness" of the innovation process was partly also the result of a deliberate innovation strategy. Large rationalisation measures were split up into individual

steps. That created a series of precedents, and the resistance which is always encountered was reduced.

72. Behind this was an "invisible network", "intuitive planning", i.e. a series of apparently coincidental but only limited aims. The steps were very pragmatic. But the aims were clear: to save staff costs, and increase the flexibility of the use of staff. Certainly the "quietness" of the innovation process simplified its realisation. The greatest difficulties appear to have been with the middle management of the specialised departments concerned.

73. Both the management and the works councils agreed that earlier consultation with the works council would have been desirable. But the management pointed out that the works council has only to be consulted on principal alterations to the organisation as such. The works council exercised deliberate restraint, as the complexity of the questions under discussion with regard to their effect on the staff could not be easily recognised in advance.

When the data sorting equipment was introduced the works council restricted itself to asking for general information and warning the specialists not to go too fast.

74. But the reality behind this peaceful and virtually unobtrusive technical change, with all its implications for reducing existing manpower and limiting future job opportunities, is that the management's strategy of "small steps" meant that each individual change that occurred was not explicitly recognised by the works council as part of a drastic overall change in the nature of the operation.

The Paper Factory

75. Works E belongs to a family owned concern, one of the leading paper producers in the Federal Republic. In 1975, 460,000 tons of paper were produced by the group, 300,000 tons of which came from Works E. Of the group's 2,300 employees, just under 1,000 are in Works E.

76. Although plans for a new stripping plant had first been tabled as long ago as 1966, the idea had been frequently shelved, mainly because the priority chosen by the company

was to expand overall production in a bouyant and growing market—particularly for newsprint—rather than labour-saving higher technology. By 1975, however, important changes in the market situation persuaded the company that it was time to implement the plan. Overall consumption of paper fell rapidly between 1973 and 1975, hitting home production and imports alike: it was time to modernise the uneconomical stripping plant, which was expensive in fuel, labour and raw material wastage.

77. With the old machinery, installed in 1939, wood was delivered, partly stripped of bark, in lengths of two metres. It had to be soaked, stripped of remaining bark and then ground into pulp. Working conditions were bad, with considerable physical strain for operatives and a high accident rate. The new plant mechanised the loading, unloading and internal transport of the wood and enabled unstripped lengths of two and one metres to be processed together. It was also able to handle relatively fresh wood, which eliminated the soaking process, giving lighter threads and thus better material for paper-making. It was calculated that the number of operatives in the stripping area could be reduced from 105 to 29. There was little experience of forward manpower planning at Works E, and this drastic reduction in manning levels gave rise to a conclusion by management that redundancies would be necessary.

78. In this instance, the traditional reluctance to involve "outsiders" in the affairs of the unit concerned was overcome, and at various stages both the Chemical Workers' Union and the group personnel management were called in. One reason for this, according to the management and the works council, was that negotiations were going on independently between the union and the Employers' Association over the revision of a clause in the collective wage contract dealing with compensation for workers transferred from one job to another, less well-paid. Clearly there was a fear that the lines might get crossed, since this was an important aspect of the introduction of the new stripping plant. Another reason offered was that specialised knowledge was needed to ensure that there were no formal errors in the new plant agreement—a not entirely plausible point, given that the chairman of the works council

has held the post for 30 years, serves as a judge in the Arbitration Court, and is a member of the industry's wage commission. In reality, given the scale of the changes, the works council probably felt that its position would be stronger if its decisions, to which it was answerable to the labour force, also had the backing of the union. In the words of the chairman of the works council himself: "If we hadn't called the union in, we would actually have finished up where we are now anyway. But we had to face our colleagues, so we got some backing. The union will have to take some of the responsibility as well now. If we had done it without the union and someone had gone to them, there might have been trouble."

79. The decision to go ahead with the new stripping plant was taken by management in November 1974, and was immediately communicated to the works council. The following month, the staff were informed at a general meeting. Construction began in July 1975, the first trial run of the new plant was in April 1976, and by 1 June 1976 the old plant was closed, the new one fully operational and surplus staff transferred to other departments.

80. Because of the initial fear that redundancies would be necessary, a halt to recruitment was imposed at the end of 1974. It is clear that what worried management was that against a background of growing unemployment, short time working and reduced overtime, the rate of voluntary departures from Works E markedly slowed down in the course of 1975. At first, the turnover rate had been encouraging: by September 29 of the 58 jobs that would be necessary to accommodate the surplus labour from the old stripping plant were already free. But in that month only three men left, and in October only one. In fact, the works council remained calm, allowing the matter to rest. By the end of the year it was apparent to both sides that no redundancies would be necessary, concentrating their efforts on negotiating compensation for men transferred. Asked why he had not intervened more strongly on the redundancy question, the works council chairman said: "That isn't the way I do things. I prefer to leave things to nature. That's always been successful so far."

81. The same relaxed style was evident when it came to discussions on the manning levels for the new plant. The attitude of the works council, reinforced by the knowledge that management had studied fifteen works where similar machinery was already operating, was to accept their figures. "We said we would wait until the plant was operating and then see what things were like. We can't do anything on spec," said the chairman of the works council. "In the next few weeks, when the plant has been running a bit, we'll have a thorough look at it to see how loaded the staff are. We may ask for one or two more to be put on." In the event, the number working on the new plant was increased by seven.

82. The main outstanding problem throughout was the maintenance of take-home pay levels for men transferred from the old stripping plant, as well as those working on the new plant, which was to operate on a two-shift rather than a four-shift system. Although the stripping plant jobs were not particularly high grade, the men were able to earn substantial supplements for night, Sunday and public holiday working. The existing wage contract provided for compensation for six months for those who had ten years' service and were aged over 45. But the contract did not stipulate the amount of compensation.

83. It was this issue that necessitated the involvement of the union and the group personnel administration, though it would appear that their role was more that of observers than participants. The agreement finally concluded dealt first with the over 45s with ten years' service who were already covered by the existing contract. Their compensatory period was extended from six to fifteen months, and they were to receive the difference between their average gross monthly pay in their old job and the gross monthly rate for the new one. All the others were to receive 100 per cent compensation, on the same formula, for the first six months after transfer, and 90 per cent for a further six months.

84. Those being transferred were offered more than one alternative choice. Noticeably, foreign workers, with their stronger orientation to earnings, opted for transfer to sections of the plant still working the four-shift system.

85. Those retained in the stripping plant were offered weekend working, at supplementary rates, to compensate for the loss of the four-shift arrangement. For the former night shift men, this meant a trade-off of higher earnings but longer working hours.

86. In summary, the introduction of the new stripping plant at Works E was conducted smoothly and swiftly, without redundancies, because of the early stage at which new recruitment was halted, and by the readiness of the labour force throughout the enterprise to accept a heavier workload during the inevitable period of staff shortage that this entailed. The readiness of the works council to accept the management's planning schedule and staffing arrangements, as well as the system of compensatory payments, was obviously a vital factor. But both sides evidently needed the reassurance offered by the presence of representatives of the union and the parent company at their discussions.

Iron and Steel Works (2)

87. Plant sections A and B belong to a steel company, which in turn is part of a large international concern with more than 76,000 employees. About 27,000 wage-earners and salaried personnel work at three plants, all located in one town in the Ruhr.

88. The company has gone through the extreme market fluctuations experienced in recent years. While the concern was still flourishing during the steel boom early in 1974, it ran into difficult times with a considerable decrease in demand up to the time of our study (April 1976). Even before that, through constant rationalisation and the introduction of new technical processes, the total workforce had been decreasing steadily for the past 15 years.

89. The works council holds a strong position in both plant sections. With the exception of 5 works council members representing salaried personnel, all of the 35 works councillors at plant section A (28 wage-earners and 7 salaried personnel) have been relieved from their normal jobs. The same applies to the 31 works councillors at plant section B

(25 wage-earners and 6 salaried personnel). The chairman of the plant section A works council has been in office for three years, his counterpart at plant section B for two years. Before that, however, he held the position of works council secretary: his predecessor remained in office for 17 years.

There is no overall works council for the entire steel company, but both plant sections are subject to the co-determination regulations of the Coal and Steel Act. The employees thus have pro rata representation on the supervisory board.

90. The flow of information from the board of management to the works council and on to the shop stewards and the workforce has been institutionalised through several boards and committees which meet regularly. Following the board of management meeting held every Monday, the labour director informs the "main committee" on the decisions taken and discussion held. Apart from the labour director, the heads of the most important central departments under the responsibility of the labour director and the works council chairmen of the plant sections and their deputies also belong to this main committee. The information thus conveyed is passed on by the deputies to the general assemblies of the individual plant section works councils. In addition to these central committees, which also include the plant committees of the individual works sections and which act as executive bodies for or within the works council, there are 22 other groups or committees in which works council representatives and delegates of the individual departments cooperate on various affairs.

91. The continuous casting plant in section A consists of two lines with a monthly production capacity of 60,000 tons (about 10 per cent of the concern's entire crude steel capacity). Line I, which became operative in January 1972, was to be supplemented as of autumn 1976 by a new, similar line. One-quarter of the concern's total crude steel production could then come from the continuous plant.

92. This construction of a continuous casting line in section A only represented the core of various measures the company planned to take. Other measures included closing down a

slabbing line in plant section B, restricting foundry operations in the oxygen steel mill of section B, constructing an automatic flame treatment line in section A, and closing down a manual flame treatment shop in section B.

93. Construction of the continuous casting line in plant section A was preceded by a careful and comprehensive examination into the feasibility of the project, including consideration of the entire production system and the current sales situation for the concern's iron and steel works. In all, 16 different versions were simulated in the planning phase.

94. Fundamentally, the crucial factor in constructing the continuous casting line was to coordinate as closely as possible the capacities of the individual plant sections within the iron and steel works. Labour market and employment considerations played only a secondary role in the decision-making process. Although difficulties in obtaining suitable labour or finding new jobs for redundant workers were anticipated, this had no real influence on the decision to construct the line and start up production at the time chosen.

95. Because of the size of the project, the supervisory board had to provide its preliminary approval, thus enabling the new constructions department to set out on their project studies and give the supervisory board the data required for a final decision. Through the discussion taking place in the supervisory board, the works council was informed of the project *formally* for the first time—the chairmen of the two plant section works council being members of the supervisory board. Informally, the works council had already been informed by the labour director, who had heard about the project earlier on in the board of management. The project was carried unanimously in the supervisory board. The works council chairmen did not make their approval conditional on promises or requirements.

96. The company had a workplace study committee meeting at regular intervals to examine the investment projects submitted by the new constructions department (in accordance with Aritcle 90 of the Works Constitution Act). This workplace study committee consisted of a representative of the company's medical service, a plant psychologist, a

safety engineer, one representative of the works council from each of the three plant sections, the assistant to the labour director, and two technicians responsible for environmental protection.

97. Members of the works council on the workplace study committee informed and consulted the plant section works councils, and also reported to the works council general assembly. In particular cases, the plant section works council consulted its shop stewards.

98. The plant operations department responsible for manning new plant and finding new jobs for people in closed-down sections, first became involved in the planning process in 1969. A manpower study was carried out in the plant sections to be closed down, showing a list of the employees and their most relevant data (age, how long with the company, current wage rating, type of job, health condition, degree of possible incapacity, job trained for, address, family status, and possible new jobs or functions). Members of the plant operations department went through this list together with the slabbing line manager in order to obtain a basic preselection. In individual discussions, and in the presence of a responsible works council member, the employees were then given proposals on their future functions and asked for their particular preferences. The employees were also given a trial period in their new jobs and were entitled to reject the offer if they found it unsatisfactory. In a few cases, employees were offered 3 or even 4 jobs in a row.

99. In April 1970 the manpower requirements committee, responsible for manpower procurement, was informed of the project for the first time. The wages and salaries committee, responsible to a large extent for determining wages at new workplaces, examined the new line for the first time in March 1971. In July 1971 the employers' and employees' representatives signed a plant agreement compiled by the wages committee and specifying the wages for the new line on a preliminary basis. The final agreement was then signed in September.

100. The employment of manpower for the new continuous casting line constituted a complex, multi-stage process. Starting up the line itself in plant section A and closing down the slabbing line in section B were only the main phases in this relatively long-term development. As, moreover, these two plant sections were distinctly separated from one another and required different personnel qualifications, they had to be treated individually as far as manpower requirements were concerned. This also applied, albeit not to the same extent, to the closure of the manual flame treatment shop in section B and the starting up of the automatic flame treatment line in section A.

101. Of the 272 employees affected either directly or indirectly by the closure of the slabbing line, 45 were able to leave the company on early retirement under the social plan.

102. The manpower policy adopted in connection with the new continuous casting line must be seen in the light of the concern's overriding employment policy.

The following factors play a role in this context:

(a) The "core team" concept. The policy is first to recruit and train people for the principal jobs on each line or unit (usually, men are recruited first). Following the responsible master mechanic, they carry basic responsibility for current operation and a high standard of qualification.
(b) The qualification procedure. Here, the policy is to build people up step-by-step for a primary job. This procedure, which usually lasts several years, also allows employees to move up in plant hierarchy.
(c) The employment stabilisation policy. The idea is to restrict the employment of new workers to a minimum and to avoid, either completely or at least as far as possible, having to dismiss anybody on account of reduced production.
(d) Maintenance of wage standards. Employees moved to a different department or line do not suffer a reduction in wages.

The employment policy shows the considerable significance of the company's internal labour market, which is always exhausted as far as possible whenever shifts and changes are to be made. The outside labour market was not of any great significance.

103. The general load exerted on the company's internal labour market at any given point in time was reduced by the policy to start up and close down the individual units step by step. Peak loads were thus avoided and the internal labour market had a longer changeover period to cope with new requirements. One difficulty inherent in this method was to get people to take over their new jobs voluntarily at the beginning, and to keep the others, who were not supposed to change over right away, on their old jobs until the old units were finally shut down.

104. Maintaining wage levels also seemed to be a fundamental problem. 24 out of the 64 workers initially employed on the continuous casting line received jobs with lower wage ratings than the ones they had had before. 22 had a higher rating, and there was no significant change for the other 18. Most of the workers demoted in this way had had top-position jobs in the old plant but were unable to reach the same position in the new one. In addition, the better working conditions provided by the new continuous casting line turned out—paradoxically—to be a problem. Workers formerly employed in the foundry pit of the oxygen steel mill had received relatively high wages on account of the adverse conditions they were exposed to. As the conditions on the new line were much better, they did not receive the same wages anymore.

105. In fact, no one suffered an immediate loss. They were protected by a basic wage agreement and a plant wage contract which guaranteed the same wages for 6 months and then a step-by-step reduction of 0.10 M/hr each month, so that the reduction process was a relatiely long one, particularly for senior workers. Nevertheless this was only an interim solution with an inevitable loss in wages as the final result.

106. The manpower policy adopted proved the efficiency of the company's system of using its internal labour resources. No major manpower problems were encountered in starting up the new line and closing down the old one. In fact, it was

not even necessary to have a general shift of workers between the two lines. Even though one of the two sections played a leading role in wildcat strikes that broke out in 1969 and 1973, and also affected the other section, these tensions do not seem to have affected the large-scale innovation described here.

107. The way co-determination was treated in the introduction of the new continuous casting line demonstrates a process of steadily increasing workforce involvement, of more institutionalised cooperation, and cumulative compilation of experience to support cooperation in general. Although this process, as the parties realise, makes the works council become more and more of a "special-purpose planning group" concerned not only with personnel affairs but also holding increased responsibility, this is not regarded as a real problem.

The Printing Industry

108. As elsewhere in the world the printing and newspaper industries of the Federal Republic were being increasingly affected by a radical technical revolution as computer-based systems of typesetting threatened to displace old methods of hot-metal setting throughout the 1970s. This in turn threatened the jobs of highly skilled and well-paid craft workers, who formed the core of their industry's union. The argument was carried through to a confrontation involving a most unusual series of strikes and lockouts in 1978, leading to a compromise achieved through mediation which was not entirely satisfactory to either side.

109. Ranged in opposition were three employers' organisations and one union. The employers' organisations represented general printing (BVD), newspapers (BDVZ) and periodicals (VDZ). The union, IG Druck und Papier, with 152,000 members is organised on an industrial basis. That is, it recruits among all print employees—journalist and clerical staff as well as skilled craftsmen. However, while membership was high among craftsmen it was low among other employees affected by the proposed changes. For example, only 16 per cent of the industry's clerical staff were

members. (Two other unions affected by data-processing technology—the German Clerical Staff Association and the Banking and Insurance Union—were also involved in the negotiations.)

110. The technical innovations were not introduced overnight, but gradually. As a result, some employers were able to negotiate the new manpower arrangements they wanted to accompany the new technology with their own domestic works councils, including the right of clerical workers to operate photo-typesetting equipment, although this covered an area which had hitherto been the preserve of skilled craftsmen. Conversely, some of the works councils were able to get the manpower agreements they wanted from their employers, including promises that the new technology would not lead to any redundancies. The fact that such agreements had already been made locally was to complicate the negotiations considerably when the technology became the subject of contention on a national scale.

111. Conscious that the position of craftsmen was being quietly eroded, the union made its first move in 1975 when it put in a claim that new photosetting equipment should be operated only by compositors (who had enjoyed exclusive rights to set print by hot-metal methods), working no more than four hours a day and at the traditional compositors' wage rate (120 per cent of the skilled wage). This claim was overtaken by a prolonged and damaging strike-cum-lockout in 1976. Although ostensibly over pay, most observers agreed that it was made more bitter by the unease felt by skilled craftsmen at the encroachments of the new technology. In any case it was not until the following year that proper discussions opened on the technology issue.

112. By September 1977, the two sides had reached an impasse. The union was sticking essentially to its view that the new terminals on which text was to be prepared should be operated by the same compositors who had worked the old hot-metal systems—or by other skilled workers where that had already been agreed—and that the old compositors' pay rate should continue to apply, while the working week was reduced to 35 hours. The employers offered few concessions. They insisted that work on the new systems was clerical in

nature and should be paid accordingly, though they did offer financial compensation to compositors who transferred to the new equipment. The union therefore invoked the statutory arbitration procedure, more, it appears, in order to clear the way for militant action than to find a solution. In the event, arbitration merely recommended a return to negotiation.

113. The union was now in something of a tactical quandary. It had been weakened financially by the 1976 strikes; some of its members had obtained better agreements through their works councils even than the union was demanding; and there were doubts as to how far the whole membership would support the minority craftsmen's interest. While gearing itself up for tougher action, therefore, the union asked works councils to refrain from negotiating new local deals, imposed an overtime ban and sanctioned local "warning" strikes. It then returned to the bargaining table and reached what the union's leaders thought an acceptable compromise: binding regulations on the operation of new equipment for five years and compensation to compositors transferred to new work on a reducing scale for six years.

114. By this time, however, the militancy of union members had outstripped that of their leaders, and the union had to return to the employers with a set of stiffer demands. (The clerical and banking unions already mentioned stuck to the compromise agreement and left the print union to fight, supported only by the Journalists' Association.) Selective strikes were ordered, and the employers responded with equally selective lockouts, but the action quickly escalated into an indefinite national lockout and worker-occupation of some plants. A compromise formula was finally worked out by mediators under the chairmanship of the President of the Federal Labour Office, Herr Stingl.

115. The agreement gave the union some of what it wanted. Compositors were to be employed on the new systems for an eight-year period, except where suitable skilled labour was not available. But the systems were classified as clerical work and rates of pay were to be tied to clerical rates. Journalists, too, were to be allowed to input text. The result was that while a fair degree of pay and job security was guaranteed for

existing compositors, the prospect was for that separate grade of skilled men to disappear through natural wastage over a period of years, their functions of typesetting being taken over progressively by journalists and clerical workers.

116. Since that agreement was made, the employers have embarked on a policy of trying to secure the elimination of all skill restrictions. This poses a dilemma for the union, which already faces the steady reduction of its core compositor membership. Should it follow militant voices and fight for the socialisation of the media—in other words, seek a political solution to its and the industry's social problems? Or should it instead seek to build a new membership in sections of the printing industry among, for example, clerical staff where the level of unionisation is only 16 per cent?

III

INNOVATION AND MANPOWER POLICY

117. In all the case studies of industrial change in West German enterprises which we undertook for this report, the express aim was to achieve higher levels of productivity and more economic use of resources. Management, technical and sales staff, as well as the members of the enterprises' works councils were all agreed on the need for technical and organisational innovation to serve this end.

In the first stages of innovation, when the managerial decisions were taken and planning began, manpower policy considerations per se usually did not play a dominant role in the considerations at least where the planned changes involved the production process. Where innovation concerned administration or support services, manpower policy aspects were more important at an early stage. In general, the case for carrying out industrial change was seen primarily as an economic or technical problem; and these were the aspects which were dealt with first. Questions concerning the availability of sufficient qualified manpower, or whether existing levels of employment could be maintained were—at least at the early planning stage—given less priority.

118. The fact that the early stages of planning were dominated by considerations other than manpower policy does not mean that this aspect did not play an important role as the project planning developed. Once the basic technical and organisational concept had been established, then the implementation of change and the consequences it would have for the enterprises' employees were closely examined. At that stage the planner sought as a rule to minimise possible

hardship that the process of change could have for the workers.

119. Until comparatively recently, the most common form of investment in German industry was to expand production. Manpower policy was, therefore, mainly concerned with filling additional jobs. The emphasis lay of necessity not so much on retraining existing employees, as on recruitment and training of fresh labour; less on redeployment of the work-force than on supplementing it from outside. Where change did have an impact on existing labour, it was comparatively easy to ensure that the individuals directly affected retained their status and income.

In fact, for the central figure in the German labour force, the skilled worker, expansion quite often meant an improvement in his professional and company status, with opportunities to acquire higher qualifications and promotion.

120. Since the oil crisis of 1974, however, there has been a marked shift away from the previous "expansive" phase of industrial innovation to "intensive" innovation, based upon more intensive use of existing capacity and higher levels of productivity. This "intensive" phase of rationalisation created a very different manpower situation. Firstly, the number of jobs has not continued to increase, but has actually decreased. Secondly, new methods of work require different qualifications, new skills, on the part of the labour force, which gives considerations such as retraining a higher importance. Dismissal sometimes unavoidably occurs and redeployment can result in loss of status and loss of income for the individual.

In confronting this new manpower scene, management had to face *new tasks.* These included:

— the selection and training of workers needed to operate the new systems;

— deciding on the future of those employees displaced by the proposed change: redeployment or dismissal;

— to minimise the impact, in terms of loss of status and income, for these workers.

121. The shift away from "expansive" to "intensive" innovation, together with the effect of the recession on the general labour market and on the internal labour markets of the enterprises, could be seen to offer management an opportunity to practise a more "aggressive" manpower policy, perhaps by using the threat of dismissal for disciplinary purposes. But we found in all the companies studied very little evidence that they had done so. In fact, there was a conscious attempt at continuity in employment policy, in particular to shield the "core labour force", i.e. mainly skilled or semi-skilled workers and white-collar employees, from the adverse consequences of the new situation. The central theme remained stability of employment and the need to avoid dismissals almost at all costs.

122. Various methods were used to avoid dismissals among the core labour force. It may not always have been possible to keep every individual in the job he had been doing, but the policy of ensuring stable employment was maintained by the adoption of such devices as:

— imposing a standstill on new recruitment;

— the introduction of short-time working;

— a cutback in overtime working, which might also mean eliminating a night shift;

— a reduction in the use of contract labour;

— carrying out work within the enterprise that had previously been done by other companies;

— early retirement of older employees;

— incentives, in the form of redundancy compensation, for individual employees to leave voluntarily.

Dismissals were carried out only if all these measures proved to be insufficient.

123. A measure of the success of this approach is that in all the enterprises studied, the slimming of the labour force was accomplished without dismissals, although considerable reductions did occur. There were, nevertheless, a substantial group of workers who, as a result of rationalisation, found

themselves down-graded to jobs which normally carried lower rates of pay. Industry-wide collective agreements, augmented by plant agreements negotiated with management by works councils, provided the maintenance of earnings for a limited period for those workers who were transferred.

Without doubt, the efforts to cushion redeployment presented enterprises with a considerable financial problem which was made more difficult by the slogan "equal pay for equal work". Employees might find themselves working with men earning higher rates of pay because of their previous job. Even so, to most enterprises the choice was clear, that when job status conflicted with the principle of stability of employment the latter was invariably given higher priority.

As far as downgrading is concerned, there is some evidence that in keeping perhaps with the tradition of pieceworking, where a worker may achieve a peak performance which declines as he gets older, such disadvantages of rationalisation seem to have been considered as an unavoidable consequence of the process of innovation.

Enterprises generally endeavoured to maintain the employment policies developed in times of full employment even though circumstances had changed, and in spite of the fact that this proved costly in a more difficult economic climate.

124. On the other hand, the enterprise quite obviously did not have to bear the full cost itself. Some of the burden was transferred to the community. Government is required to foot part of the bill for compensation when short-time working and early retirement is required.

In other less direct ways the cost is transferred when a ban on recruitment puts a company's internal labour market in a privileged position at the expense of the general labour market, representing lost employment opportunities and a consequent growth in unemployment.

Further, part of the cost was transferred to an enterprise's own workforce, when workers had to transfer to lower grades of work at lower rates of pay, overtime earnings were cut, and short-time working meant smaller wage packets. Also more output was demanded of the individual worker with faster machine speeds on productivity schemes. Although such arrangements were normally defined as temporary, we suspect that some of them will remain as a permanent feature.

125. There are other implications for employment policy. There is, in particular, a loss of flexibility in personnel policy. Having opted for stable employment, management could not, or, as we pointed out earlier, did not, use the changed labour market situation to impose tighter discipline. Options for further transfers of workers to different jobs within the plant are limited, labour turnover becomes much lower, and mangement may be forced to extend the advantages once reserved to the core labour force to peripheral groups of workers.

126. Faced with this loss of manoeuvrability, management had developed an unobstructive way of preserving a margin of flexibility in manpower policy. This might be termed "silent" employment, working alongside the "official" employment, policy with its legal constraints and its key objective of stability of employment.

An important feature of "silent" employment policy is its low-key approach. "Exemplary" dismissals, for example, are avoided. The law relating to redundancies and dismissals is punctiliously observed. This legal framework appears, however, to be flexible, and this is due in large measure to the posture adopted by works councils. The legality of a dismissal on disciplinary grounds frequently depends on whether it is acceptable to the works council. In other words, the "silent" employment policy depends heavily on their cooperation, and it seems that this cooperation is forthcoming over a wide range of issues. This is particularly evident where the interests of the majority of employees—the core workforce—do not coincide with those of the fringe groups.

The importance and the extent of "silent" employment policy must not, however, be overestimated. Even so, it exists, and it acts as a corrective to "official" employment policy which, of course, remains quantitatively far more important. It is a mechnism of adjustment enabling the enterprise to live with a rather restrictive legal framework by allowing management some lee-way.

127. Another effect is a more carefully planned, longer term personnel policy. Its aim is the more intensive use of human resources within the enterprise, to ensure greater independence from the outside labour market. This has become more important with the concentration on "intensive" innovation, one result of which has been a tendency to meet labour requirements from within the enterprise. However, this may well create new problems for the external labour market.

IV

INNOVATION AND INDUSTRIAL RELATIONS

128. Rationalisation within an enterprise constitutes an area of potential conflict between management and workers' representatives. It imposes strains on the familiar and well-tried system of "give-and-take" which normally operates, throws established methods into question and requires that the components within the system should be defined in a new way. The issues being dealt with constitute traditional areas of contention between management and workers' representatives.

It is, therefore, perhaps surprising that in all but one of the cases we studied the planning and implementation of industrial change was almost free of conflict and was regarded by neither side as controversial. Though it was sometimes negotiated vigorously, it did not lead to open conflict or confrontation. For the most part, rationalisation schemes appear to have been conceived and carried through in an atmosphere of neutrality.

In this context, the case of the printing industry appears atypical. Several reasons may account for this. Printing is the most ancinet of industrialised processes and in every industrial country its employees have developed correspondingly venerable systems of craft protection which have in turn produced a high level of incomes. The new technology promised radically to deskill the old, highly paid trades, which would in future become accessible to relatively untrained workers willing to accept lower wages. And while the craftsmen formed the core of the union (as of the industry) they did not form a majority of union members, so the protection of their interests was neither in the interest of the industry nor of the union majority. Finally, the introduction of new technology began at enterprise level where agreements set conflicting precedents—some more, some less, favourable to the craftsmen—when the argument reached the national level.

129. In general both sides were basically in agreement on the goals that change was designed to achieve:

— the basic necessity for, and desirability of, rationalisation was not questioned. The issue was not whether but how change should be implemented;

— the aim of increasing productivity was undisputed;

— a reduction of the workforce was seen as a possible consequence of rationalisation, as was the possibility that some workers might suffer a loss of status;

— there was unanimity on the necessity to reduce any adverse impact on the workforce as far as possible;

— it was acknowledged that where individuals were adversely affected, they must receive financial compensation.

130. In the enterprises we studied, the works councils were generally informed of impending change at an early stage. Very often they knew of the project before the main company board and the supervisory board had given them their official blessing. Quite often early information was passed on to the works council informally, e.g. by discussion between the chairman of the board and the chairman of the works council. In those enterprises where co-determination law operates, the works council was informed by the labour director who, as a member of the board, was able to give detailed information. This ensured the participation of the works council in the decision-making process.

The situation was less satisfactory in those undertakings where the initiation and planning of a project were carried out not at plant level but centrally, i.e. at the headquarters of a concern. In these cases, the local works council was often presented with a fait accompli, and the same seems to have applied to the plant personnel departments of the relevant plant. Thus, the role of the works council was restricted to agreement or rejection of the proposals put before it, though it still had the power to insist on certain conditions being met; for example, that there should be no redundancies.

In the single case study involving change throughout a whole industry—printing—the works councils played a somewhat different role. In the early stages of technical innovation some works councils agreed to accept and work equipment whose introduction was later resisted by other groups, while other works councils secured job preservation arrangements to which employers in the industry as a whole were unwilling to subscribe. The standards set in plant level negotiations, whether they favoured employers or workers, thus inevitably set standards which both sides sought to make general when the agreement reached the national level.

131. Once a project had been officially approved by the main and supervisory boards, implementation varied considerably from plant to plant. In some cases, the works council took a prominent part both in ensuring that legal requirements were met, and in submitting proposals for ways in which the project could be improved. This was not necessarily confined to traditional social affairs—questions of recreation and washroom facilities—but the physical environment such as the planning of buildings and the siting of machines, which took them deep into the realms of technical and organisational planning. On the other hand, there were enterprises where, although the works council was informed of projected change, it took little further part in the way such change was executed.

132. Summing up, it became clear that during the stage of initiation and planning of a new project it was the technical and economic departments which dominated the process of change. As the plans were implemented, however, works councils and personnel management were increasingly involved.

The question of what weight can be attached to the influence of the works council on a project is difficult to assess. Firstly, we must emphasise that it is less the moment that the works council is informed of the project than the degree of continuity with which they are involved and the institutional arrangements for its involvement which vary.

From this we may deduce that there were considerable differences in the actual influence exerted by the works councils from plant to plant.

In fact, where the involvement of the works council was not institutionalised much depended upon the initiative and the vigour of individual works councils.

133. In none of the individual enterprises studied did negotiations between management and works councils result in major delays to a project. Although the works council did not question the basic need for the project, its agreement—which again was forthcoming without significant delays—was often subject to certain conditions being met. Among these conditions were:

- that dismissals should be avoided, and that workers who left voluntarily should receive adequate redundancy pay;
- that workers transferred to other jobs within the plant should continue for a limited period to receive their old rates of pay;
- that works council representatives should be present when staff transfers were being considered;
- that the works council should be involved in the organisation of social facilities, such as canteens, washrooms and recreation rooms;
- that all legal requirements concerning the working environment and safety at work were observed;
- that the works council should be consulted on wage scales for newly created jobs.

Generally, works councils' assertions of rights in all these areas seem to have been accepted by management as their legitimate concern. At least, we were told of no disputes about the legitimacy of works council intervention.

V

COOPERATIVE CONFLICT RESOLUTION

134. Rationalisation, in spite of its inherent potential for conflict, has been achieved in the past largely without confrontation or excessive delays, thanks partly to the series of favourable economic conditions outlined earlier, which helped defuse possible areas of conflict. Clearly, it was easier for change to be secured during that "expansive" phase of industrial investment in the Federal Republic with its accompanying period of full employment in an exapnding economy. As we have shown, these conditions have changed in the past few years. It was, therefore, somewhat surprising to record that little hardening of attitudes appears to have occurred in the enterprises studied, even when both the character of technical/organisational innovation and the general labour situation have made things more difficult. (Again with the exception of the printing industry, as described in paragraph 128.) Indeed, we were impressed by the resilience with which the industrial relations systems within these enterprises coped with the problems of rationalisation under these new conditions.

135. Any attempt to explain the resilience of the system should be placed in a wider context. It can only be understood if it is seen as being embedded in a style of industrial relations which we have chosen to call "*cooperative conflict resolution*". This style has the following characteristics:

— the concern of both parties to avoid open confrontation which forces one side to take up an attitude from which it cannot retreat without loss of face;

— a willingness to seek compromises which respect the interests of both sides;

— renunciation of the chance to exploit to the extreme the legal position or bargaining power possessed, on the understanding that by not adopting a maximalist position, the other side will reciprocate when their position is the stronger;

— while the well-being of the enterprise and its workforce is regarded as having precedence over any sectional interest of the negotiating parties, acknowledgement at the same time that these interests may well diverge;

— a tacit understanding that while the legal framework within which the negotiation takes place is binding, solutions must take account of the special needs and conditions of the enterprise. This, typically, would mean that the two sides reach an informal agreement in line with the requirements of the enterprise without reaching the point at which the letter of the law must be invoked.

Having listed these characteristics of the cooperative conflict resolution system, we should not look at the way the system works and how these characteristics are displayed.

136. Although provisions defined by law or collective bargaining agreements—and here the existence of the works council law is crucially important—influence the positions adopted by the negotiating parties, they do not prescribe the way negotiations are conducted in individual cases. Rather, they act as a general orientation for action. The existence of the legal framework, which is accepted as binding on both sides, is a basic precondition for the flexibility which is exercised in negotiation. It is taken for granted that the law and collective agreements will be adhered to. There is also, however, an understanding that, although the law exists it does not have to be constantly invoked to the letter.

This flexibility, amounting almost to nonchalance, is made possible by an informal code of behaviour for the bargainers. Thus, each side knows what it can expect of the other, and these informal rules of the game have produced an attitude of mutual trust and cooperation.

137. Another important element of the system is the fact that the demands made by each side are legitimised not simply by reference to the works council law, but also because each side makes clear that its own interests coincide with those of the enterprise. Both try to present their case in a form which stresses its advantage to the enterprise. This makes it difficult for the other side—equally commited to the well-being of the enterprise—to abandon its cooperative attitude.

There is no doubt that each side has to develop a sensitivity about how far the other will go to accommodate its demands. Only against this background can it be understood why most committees and working parties do not take formal votes when they meet. Instead, points are discussed until a general concensus is reached.

138. Cooperative conflict resolution tries to solve all problems within the confines of the plant, i.e. negotiations between plant management and works council. Points at issue between the parties ususally concern conditions within the enterprise and are, as a matter of principle, solved within the enterprise. Calling in help from outside institutions is regarded as failure and is avoided, if at all possible. It, therefore, appears that the works councils do not generally regard the relevant union as a court of appeal if agreement with management becomes impossible. The trade unions are instead looked upon as a body which can provide facilities not available at plant level, e.g. training.

The introduction of change either from within the enterprise or from outside (e.g. a change in the law) brings with it the possible danger that, as a result, the parties will be drawn into open conflict, that intransigent positions will have to be adopted by both sides during the course of negotiations as a consequence of which a solution will be found only by relying on the letter of the law. One of the main contributions of cooperative conflict resolution is to prevent the parties from being forced into or stumbling into open confrontation. Both management and works council executive spend a good deal of their time trying to avoid such danger of open confrontation. Each side is determined to keep things cool and, if possible, to prevent conflict from arising at all. It is only in this context that the frequency and continuity of contacts between both sides can be understood. In the enterprises we studied it was found that the works council chairman and his closest colleagues were in continuous daily contact with works and personnel management.

139. An important aspect of these frequent contacts is that they serve as a clearing-house, where each side informally explores the position of the other. Each then gets an indication of how far they will be able to go when formal negotiations commence. By offering hints of their own attitudes, the soundings often produce an outline of the compromise which will be reached or ratified at the formal negotiation. Again, at the formal meetings the two sides may well confine themselves to what has been discussed informally as part of the "preventive" strategy, re-enacting what has already been agreed behind the scenes. By this means the risk of failure, conflict or breakdown is avoided.

In a similar way, another recognisable feature in the working of cooperative conflict resolution is the concept of "anticipatory conflict resolution". This, as the term implies, involves the identification of potential areas of conflict well ahead, thus structuring and neutralising any danger that such topics might hold. One example of this was the way in which a number of large enterprises anticipated the 1972 legislation which greatly increased the power and functions of works councils, by phasing in well ahead of the law the reforms that were to be enacted. In this way, any possibility that the enterprises would appear to be acting under pressure was avoided. Plant level collective agreements also operate as an instrument of anticipatory conflict reduction, where agreements may be drawn up to meet circumstances that have not yet arisen, or have been the subject of formal claims by employee representatives. One example of this type of action is the agreement reached, during a period of full employment, to cope fairly with mass lay-offs. As one personnel manager expressed it, "The plant agreement we draw up in good times makes life easier in bad times".

140. The informality of the approach, the importance of unwritten ground rules and the continuing process of conflict avoidance all mean that cooperative conflict resolution is particularly dependent on the personal qualities of the participants on each side. Their individual attitudes and prejudices play an important part in the discussion. Only those who are particularly alive to the nuances of this style of procedure can practise it successfully.

The representative of the other side does not appear as "the enemy", but as a fellow player in a game which demands from time to time that the participants adopt opposing attitudes. Nevertheless, it is accepted that each side needs the other if the game is to function.

141. If cooperative conflict resolution is to run smoothly, it is important that there should be continuity in representation on both sides. Only where most of the participants are thoroughly familiar with the "rules of the game" as a result of many years' practice, can one expect that the precarious path between confrontation and the form of cooperation, as described above, can be navigated.

142. The participants in the process of cooperative conflict resolution must present themselves to their own constituents as staunch defenders of their interests. This applies in particular to the works council members which, as the elected representatives of the workforce, must demonstrate their success as effective and determined negotiators. They are, after all, interested in being re-elected, and must also show the relevant trades union that they are an active and influential body, representing workers' interest.

Management must have understanding for the particular needs of the works council representatives to achieve "results". This must be done in a form which shows that concessions have been "wrung" from the other side. In this way, agreements, which have been reached relatively easily, are sometimes dramatised when they are presented publicly.

143. We have described the basic principles of cooperative conflict resolution. We have not yet clarified why it has proved such a resilient system, particularly in relation to rationalisation. Our contention is that both parties, while

pursuing their own interests, are bringing certain expectations to bear, which make their involvement worthwhile, but also represent the basis for a system of rewards and sanctions which supports and stabilises it. And by definition, such a system is able to exert pressure not by the exercise of its sanctions, but merely by the threat of doing so. Let us look at some of the weapons available to each side of the negotiating system:

— Management has at its command a range of material and financial benefits for labour force going beyond its commitments in legally binding collective agreements. These can be "released" to demonstrate that the works council is doing a good job in negotiating for the workforce. Thus, we have a reward or gratification that can be used or withheld by management in its relations with the works council.

— Another sanction-gratification available to management is to furnish the works council with offices, secretarial assistance, cars and so on. This doubtless increases the efficiency of the works council, but it carries the implicit threat of removal.

— Again, while management is bound by law to supply certain information about the enterprise to the works council, it may go well beyond the statutory requirement—a gratification.

— In co-determination, management may open up certain extra areas of influence to the works council—again it is something that can be taken away again.

Much the sample applies to the works councils. They also have command of a system of rewards and sanctions, of which the following are some examples.

— They can either make a fuss or keep quiet over issues where from a legal standpoint they would be entitled to intervene on behalf of the workforce. In other words, they can choose to press or not to press a legal advantage.

— Works councils are sometimes asked to take over disagreeable managerial functions—e.g. to select workers for transfer and downgrading or for dismissal. They could refuse or accept.

— Works councils sometimes share responsibility with management for certain decisions which would otherwise be difficult to persuade the workers to accept—e.g. technological change. So another sanction is available to them.

— Works councils neutralise and absorb some of the pressure from the labour force that would otherwise have a direct impact on management. There is a sanction-reward dimension to a works council sharing responsibility for management's disciplinary decisions.

144. Cooperative conflict resolution can thus be seen as a system of interdependencies in which each side in looking after its own interests is dependent on the cooperation of the other side. That means that they have the power, or the potential, either to grant or to withhold this cooperation. Both sides, in other words, have the means to enforce good behaviour by the other.

145. What is the price that has to be paid for the good behaviour of the other side? Is it in reality that a balance of interests is created in which neither side is capable of conflict any more? That would imply that the system itself involves a collective suppression of conflict, and the suggestion that some basic interests of enterprise or workers may be neglected by the deadlock produced by the complicity of their respective representatives in management-works council negotiations. In particular, the works council could, in demonstrating so much loyalty to the cause of the enterprise, run into danger of becoming incapable—should it prove necessary—to enter into confrontation with management.

However, the evidence of our case studies seems to suggest that a complete suppression of conflict did not occur. There are a number of objective reasons why this is so.

146. The works council must be concerned to justify its actions to the workforce it represents. For this reason alone, the works council is obliged to pursue policies which, in a limited way, forces it to challenge management.

Further, it is inherent in the logic of cooperative conflict resolution that the works council should use that margin of influence given it by the legal framework and power realities of the situation. Only then will its willingness to "behave well" carry weight with management. If the works council fails to use this margin of influence in its dealings with management, the usefulness of its willingness to cooperate would be diminished. The greater the weight of cooperation, the more effective is the threat to withdraw it. Therefore, the active use of the whole field of influence opened up by co-determination implies not only co-responsibility but also the entrenchment of the works council's position.

That is due in no small measure to management's interest in having an "active" works council; only a works council which is able to present itself to the workforce as independent can succesfully fulfil its role as a disciplinary organ. An inactive or obviously "bought" works council would be useless in that role.

147. However, it is not impossible for management and works council between them to produce that degree of minimum confrontation that convinces the workforce that the works council is not a pliant tool of management. In our discussions with both sides, we were assured that such manipulation was not possible and that one was wary of such publicity, as it did not pay. Though each side had to be prepared to compromise, its own interests had to be borne in mind. For this reason, there were clear limits to such a game of mutual complicity. Sooner or later, each party had to chalk up definite successes of its own.

There can be no doubt that these safeguards help protect the system of cooperative conflict resolution against corruption. It may also explain why this system in the enterprises we studied has surmounted the strains of the last few years relatively well—without the power of the works council having been curbed, and without open conflict breaking out.

148. What have been the results of the exercise of cooperative conflict resolution for management, for works councils, for the enterprise and for the workforce?

In those cases where management and works council have played the game according to the rules—in other words, where a balance of cooperation and confrontation has been maintained—the effect has been to protect both management and works councils against their constituents, i.e. the owners and the workforce. Their operation of the system is confirmed by the regular re-election of the works council, and the increased weight within the management structure of the personnel department.

As far as the enterprise is concerned, one of the central contributions of cooperative conflict resolution has been to guarantee it a certain measure of autonomy against outside influences by providing an "in-house" vehicle for stabilising industrial relations, and reducing the probability of uncontrolled conflict. Both sides thus have an efficient negotiating platform for the airing of grievances, for the making of concessions and for articulating their expectations. And once agreement has been reached, it can be implemented immediately.

Cooperative conflict resolution gives the enterprise a certain measure of flexibility and freedom within the framework of external rules and forces that govern West German industrial life—the workings of the labour market, safety and health regulations, industrial relations laws, collective agreements, and so on.

The execution of technical and organisational change is facilitated by cooperative conflict resolution, which simultaneously ensures the adoption of manpower policies that reduce the element of controversy from industrial change, and so help to expedite it. It also plays a central part in the process we have already described above, which enables the enterprise to transfer part of the "social" cost of rationalisation to the community, via the social insurance system (e.g. early retirement) and the general labour market (e.g. a stop on recruitment).

So, for the individual enterprise, cooperative conflict resolution has undoubtedly been a "success". It is more difficult to assess the effect it has had on the workforce. For the core labour force, i.e. mainly the skilled workers and the white-collar employees, the manpower policy that has been influenced and guided by cooperative conflict resolution in the large plants has had substantial advantages. They, after all, are the predominant group represented in the works councils, which naturally look after their interests. Nevertheless, a price has had to be paid for these advantages, notably in the increased intensity and pace of work, in the increasing division of labour, and in the consequent devaluation of existing qualifications. These adverse aspects of the system at first seemed not to weigh too heavily since their impact was gradual, and because they were felt mainly not by the core workforce but by the fringe groups—the unskilled workers, the women, the foreign workers. The situation of these groups may well have deteriorated.

149. In this context, we have to consider the disciplinary role of the works councils and their participation in the "silent" employment policy. In this, the works council acts as the representative of the strong core workforce against the weak fringe groups and against deviating individuals. That makes them open to the criticism that they are more concerned with securing the privileges of the strong than with improving the lot of the weak.

150. Cooperative conflict resolution is only one of several styles of contention between works councils and management. We have so far concentrated on it because it was the predominant style in the enterprises we studied. But we must not overlook the fact that our selection of case studies was mainly restricted to the successful—that is, to enterprises where industrial change was carried out without conflict. From other studies we know that there are other styles of contention. We intend, therefore, to describe some of these now, not so much as to give a detailed description, but to put the style of cooperative conflict resolution in perspective.

Open Confrontation

151. In this style, management and works council regard negotiations as a struggle between basically conflicting interests. The interrelationship between the interests of the

labour force and the interests of the enterprise is perceived, but is not regarded as the guiding principle behind every action (in cooperative conflict resolution, it will be recalled, the objective of the negotiators is to identify their case as closely as possible with the interests of the enterprise). In the open confrontation style each side defines its position in opposition to the other, and demands are formulated and justified in terms of the sectional interests of one side. Maximalist positions are adopted, with the emphasis on legal rights, national agreements, the exploitation of the power situation in terms of the labour market, the company's sales position, etc., and the willingness of the labour force to resist.

Accordingly, the regulations of the Works Council Law assume a great importance. Negotiators arm themselves with a thorough knowledge of the legal framework and learn to use that knowledge to back up and reinforce their demands. They aim to negotiate extensive formal plant agreements. Management views the works council mainly as the source of demands. The works council resists management attempts to allot it a role in the disciplinary system, where it might be seen as acting against the labour force. Personal contact between the two sides is limited in the main to the work of commissions and committees, where the works council will insist on numerical parity, votes will be taken, and decisions formalised, preferably in writing. The element of mutual trust, confidentiality and informality, together with the observance of the unwritten rules of behaviour, which characterise the cooperative style, are missing.

Suppression of Conflict

152. In this style the works council acts essentially as an extension of management, accepting that the well-being of the enterprise is the overriding and highest interest, coinciding to a high degree with the interests of the labour force. As a consequence, the idea of an independently defined workforce-interest almost disappears. The works council does not confront management with a constant series of demands. Instead, it sees its role as delivering what management expects from its employees, in particular from fringe groups such as foreign workers and newly employed workers. In

fact, the works council takes over certain tasks from management, supervising disciplinary and work procedures, time-keeping, the implementation of redundancy measures, and the distribution and allocation of social benefits.

153. It will be apparent from the brief description of these contrasting styles of contention that cooperative conflict resolution represents a middle way between them. In attempting to reconcile the interests of the workforce, this style tries to steer a narrow path between open conflict on the one hand, and total suppression of conflict on the other.

154. We would be guessing if we tried to assess the quantitative distribution of these different styles of contention among German enterprises. The few cases in our survey did not provide a sufficient basis for assessment, all the more so since we confined ourselves on the whole to successful instances of innovation and change. The results of other studies, however, would lead us to assume that in manufacturing industry, particularly among middle- and large-scale companies, the cooperative conflict resolution style predominates. On the other hand, we have no evidence concerning its adoption in administration, in service industries, in smaller enterprises, or in the headquarters of large-scale concerns.

155. We are conscious that in our description of the cooperative conflict resolution styles, even in those enterprises which were the subject of the case studies, we were dealing with only part of the whole field of conflict resolution. What we have described is a relationship between management and the institutionalised representation of the labour force through works councils at plant level, or to be more precise, of an executive elite of the works councils consisting of the chairman and one or two of his closest colleagues. It cannot be excluded, and there are indications that point in that direction, that other areas within these same enterprises might be quite different: for example, if we had looked at the relationship between workers and supervisors. Again, examination of the behaviour and attitudes of other members of works councils, who do not belong to the executive elite, and shop stewards, might demonstrate a different orientation

to what we found among the works council "establishment". What is evident, however, is that no pressures, impulses or disrupting influences have emerged so far from these other areas decisively to change the cooperative style that has been developed.

All these reservations lead us to the conclusion that with the cooperative conflict resolution style we have described only one part and only one aspect of the reality of industrial relations in the Federal Republic. We have concentrated on it, however, because it has set a pattern for contention between management and workers at plant level which is at the disposal of, and has to an important degree shaped, industrial relations in large parts of Germany industry. In particular we believe it has influenced the inter-relationship between manpower policies, industrial relations and the process of rationalisation within the enterprise—in a way that we believe is rather unique to the German situation.

VI

THE EXTERNAL FRAMEWORK

156. So far, we have mainly concerned ourselves with the relationship between cooperative conflict resolution and rationalisation within the enterprise. Now we should consider the influence that external forces have exerted over this system.

It is first necessary to stress the importance of the legal, institutional and collective bargaining framework which regulates German industrial relations.

The success of cooperative conflict resolution in handling industrial change owes much to the legal framework. In particular, we must look at the *Works Council and Co-determination Laws* which govern the representation of workers' interests at plant level.

As we pointed out earlier, there is a distinct division of labour between the trade unions and the works councils which confine the unions mainly to national affairs outside the enterprise. The enterprise is the domain of the works councils, and the trade unions are allotted only a guest role. Thus, management has to deal with a body, which draws its strength and legitimacy from the workforce who vote it into office, and which both thinks and acts in terms of the enterprise. Thus, one of the essential preconditions we have identified in making cooperative conflict resolution possible is established: the willingness to confine the area of conflict and its resolution to the enterprise.

The existence of the law is vital to the system, for it provides necessary parameters within which management and works council can play out their roles, which in turn allow them to behave with relative ease and magnanimity, making concessions where expedient. One result of this is an absence of haggling over the interpretation of the law in the enterprises we have studied.

It could be said that the style of cooperative conflict resolution is in a sense a response to the law which management and works councils have developed between them. The law is there, and it is binding, but it is also found to be possible to adjust its operation to the exigencies of life at the plant or enterprise. This implies that the Law only acquires its specific reality when it is processed through the workings of cooperative conflict resolution.

The statutory duty the Law lays on works councils to "cooperate in a spirit of trust" and to respect the interests of the enterprise, as well as the fact that works councils are precluded from initiating strike action, would seem to suggest that they are deprived of sanctions they could use in support of their bargaining objectives. And that, in turn, suggests a certain asymmetry in the system which loads it against the workers' interests. There is, however, an important factor which helps to offset the one-sidedness of the Law: the power of the works council to cooperate, and to withdraw cooperation.

157. The importance of such cooperation on the part of the works council is further emphasised by the existence of a comprehensive system of legal enactments and binding collective agreements which lay down minimum standards for the protection of workers economically, socially and environmentally. The fact that these usually define only minimum standards means that there is a margin for improvements that provides a wide range of items that can be the subject of negotiation. Such negotiable issues constitute an important element of manoeuvre, of give and take, which is "processed" within the cooperative conflict resolution system and in turn gives added importance to it.

So cooperative resolution depends not only on the existence of the Works Council Law, but also on the buttressing framework of statutory minimum conditions and the provisions of collective agreements, which in law the works council has the duty to supervise, or "police". This means that, in turn, the Works Council Law itself assumes its meaning or reality from this supporting framework, thus emphasising the interlocking nature of the system.

158. There is another aspect to the influence of legal requirements and nationally negotiated collective agreements. The very existence of a network of legal entitlements which regulate working conditions at plant level imposes on the works council both a controlling function—for it has a legal duty to ensure that minimum standards are observed—and a reactive function. The sheer volume of obligtions, laid down for the works council to fulfil, diminishes tendencies for it to become an initiating body, for example, by campaigning for better working conditions which are already laid down. Thus, the works council is less likely to adopted the inherently more conflict-prone role of initiator.

159. To complete the catalogue of external influences, it can be observed that the style of contention between employers and trade unions at national level is reflected in the way that management and works councils approach their differences. The German system of collective bargaining reduces the probability of open conflict. Its built-in mechanisms include a legal "commitment to peace" and a strict limitation on the opportunities for open warfare to limited periods of time and then only for certain stipulated reasons. Secondly, negotiating procedures are determined by statutes fixed by collective agreements and by the constitutions of the trade unions. A complex multi-stage process ensures that each party can press its demands only gradually. A series of predesignated steps inhibits any rapid escalation into open conflict, or the danger that either side will accidentally stumble into confrontation. At every stage, there is opportunity for reflection, revision or retraction in an intricate minuet of gradually exerted pressure that itself fulfils some of the functions of open conflict, while rarely going over the brink.

This relative absence of open industrial conflict at national level means that a backwash of unrest is less likely to develop at plant level, and the disturbance effect of the external negotiations taking place is minimal. The same is not altogether true in the context of national negotiations, which do in one sense carry the potential for localised conflict. This possibility arises because collective agreements between employers' associations and trade unions tend to be geared to the capacity of the weaker, less productive enterprises. There is, therefore, once again a negotiable margin of improvement at which negotiations within the stronger enterprises can aim. As we have noted, cooperative conflict resolution is strengthened when it is given an extra bargaining dimension.

160. The second important feature of the national bargaining system in the Federal Republic, as far as enterprise industrial relations are concerned, is the trade union structure, which follows the principle of industrial unionism. Thus the employees of each enterprise—with the exception of some white-collar staff—are represented by one trade union. As a result, demarcation disputes and interunion competition are almost unknown. This, of course, also means that a solid majority of the members of the works council will belong to the same union, thus giving it a confident and clearly defined relationship with the appropriate union. All this contributes to the scope that effective works councils and management have of maneouvre, compromise, concession and, where necessary, sanctions. The national system thus acts as a support for cooperative conflict resolution at enterprise level.

161. In assessing the importance of external factors for the development of cooperative conflict resolution, we have finally to look at the social and economic situation in the Federal Republic. Until the oil crisis of 1973, the stability of the industrial relations system in Germany contributed to economic expansion. It would be equally true to say that economic expansion contributed to stability in industrial relations. Not only did economic prosperity over a long period guarantee full employment, but it also brought considerable growth in the size of the workforce. Enterprises were able to pursue stable employment policies, while getting on with "expansive" investment programmes, rationalisation, innovation and change. Improved training and retraining facilities, together with promotion opportunities benefited the core labour force. The consequential rise in productivity ensured industry that competitive edge it enjoyed for many years, while creating the economic basis for its successful

manpower policy. Unpleasant side-effects were in the main confined to weak peripheral groups, such as foreign workers, women and the unskilled.

162. As a consequence, the trade unions were able to adopt a cooperative attitude towards industrial rationalisation and measures to increase productivity and implement new methods. There was no question of slowing down change: the constant aim was to keep the adverse consequences for the workforce to a minimum. Such cooperative policy could be maintained because the workers were guaranteed both security of employment and an increase of earnings. As a result the increased spending power of the workers gave additional impetus to economic expansion, enterprise manpower policies could come up to union expectations and discontent on the part of the workforce be prevented.

163. All these factors taken together created a climate of predictability, allowing long-term planning and innovation to be conducted with apparent ease. Rationalisation schemes hardly ever had to be abandoned or delayed because of fears of possible conflict between management and the leadership of the workforce. The lack of controversy surrounding industrial change can, therefore, be seen as the result of a constellation of favourable circumstances, the existence of which was in turn continuously perpetuated and consolidated by it.

164. However, two key elements in that constellation, namely full employment and general expansion, suffered a setback as a result of the oil crisis of 1974. After that, there were periods when inflation ran at a higher rate than the growth in incomes. Important basic preconditions which had contributed to the success and stability of employment policy over two decades were suddenly thrown into question. So for the first time, the system of cooperative conflict resolution, developed during the days of prosperity, was being tested in an adverse economic situation. Would it be capable of proving its effectiveness under these less favourable circumstances, or would it be revealed as a "fair weather system"?

Our case studies demonstrate that in the main cooperative

conflict resolution has proved surprisingly adaptable and resilient. We found that the system had developed a remarkable capacity to overcome, at least on a temporary basis, new and difficult circumstances.

BRITAIN AND WEST GERMANY COMPARED

I

THE BACKGROUND OF CHANGE

1. We now bring together for purposes of comparison the results of our experience in studying industrial change at enterprise level in Britain and Germany and the industrial relations systems of each country through which that change is achieved. In this chapter we compare the evidence of the two national reports, which were based on case studies of large enterprises. For the most part, change was successfully carried through in the plants that were investigated, although in one or two instances plans were altered, or remained uncompleted at the time enquiries were concluded. The evidence thus acquired inevitably gives only a part of the picture, and it has been set as far as possible in the wider context of the economic and industrial background that obtains in each country. First, we look at the aims, purposes and methods of the enterprises which have been subject of the case studies undertaken for the national reports. Next, we describe and compare the institutions through which employees and managements resolve their differences and the external influences that are brought to bear on their actions and attitudes, particularly those of government. Then we consider the costs which industrial change imposes on the community as a whole.

2. But before we go on to these detailed comparisons, we wish to suggest a broad theme arising from the general reputations of the industrial relations systems of Britain and Germany. In Britain, industrial affairs are carried out against a background cacophony of press headlines, sombre television coverage and a publicity glare that sometimes almost blinds the participants; while by contrast Germany

conveys the impression of a well-ordered, smooth-running, almost trouble-free system which, when it occasionally erupts into confrontation over pay or technical change or the further extension of worker participation, surprises observers. Our investigations suggest that these impressions contain a good deal of significance. While there are, of course, underlying similarities between the two countries—as there must be between any two advanced industrial nations—we have been struck by what we term the differing styles for resolving conflict adopted by each country both nationally and at enterprise level. In Germany, the prevailing style is essentially cooperative while in Britain it takes an adversary form. In Germany, employees and managements address the problems of industrial change together, expecting to reach a compromise by negotiation through a set of agreed and predictable rules. In Britain the same sort of problems are dealt with through collective bargaining procedures which follow no uniform set of rules and which therefore carry the danger of open dispute. These descriptions are not definitive. Cooperative relations exist in Britain, just as contention exists in Germany. Nevertheless, these divergent styles—cooperative and adversary—emerge as strongly characteristic of the two countries.

II

AIMS AND METHODS

3. The overriding aim of the changes initiated by enterprises covered by the case studies was in each case a reduction in labour costs. Reasons for the adoption of this aim varied widely: an overnight halving of the market for a company's products; the introduction of radically new forms of technology; the exhaustion of a resource such as coal at one site and the expansion of production at another; or, quite simply, the long-term and gradual attempt to make a better overall use of resources of capital and labour.

4. But, while the aim of reducing labour costs was common to British and German enterprise, the means chosen differed sharply between the two countries. In Britain, the solution almost invariably chosen was an absolute reduction in the size of the labour force, usually through dismissals. In Germany, on the other hand, a reduction in the workforce was also sought, but by long-term phasing, early retirement, a halt to recruitment and the redeployment of manpower within the enterprise, and thus dismissals were usually avoided. In only one of the German studies, a colliery, were employees actually dismissed from the company. But in Britain only two enterprises gave undertakings not to dismiss employees who did not want to leave, although these undertakings rang a little hollow since neither enterprise had plans fully to employ its entire workforce in productive activity after the changes envisaged had been carried through. The undertakings were, in effect, gambles, made to try and win the confidence of employees and so avoid open confrontation; management's real plans were to slim down its establishment by such inducements as schemes for early retirement or the

preservation of the incomes of workers who left at their old high levels. In a third enterprise, efforts were consciously made to develop alternative work for displaced employees by financing factories to bring new businesses to areas hit by redundancies. As yet, this scheme has produced very limited results. Elsewhere among the British cases, the impact of dismissal ranged across a broad spectrum, from the sacking of half the manual workforce in a matter of days from the first announcement, to the shedding of a comparatively small percentage over a number of years with full consultation and largely by way of natural wastage.

5. There appears to be sufficient evidence in the case studies to support the proposition that British and German managements take very different attitudes towards employment policy, so far as that involves stability of employment. For German managements such stability is a conscious objective: for British managements it is certainly desirable, but it is not a priority. This has probably tended to inculcate among British workers a more acute sense of anxiety about the employment prospects than among German, and this in turn is likely to have adversely influenced the attitudes of British workers to any sort of change which carries with it the possibility of job losses.

6. A number of reasons for this divergence suggest themselves. Until recently, the German labour market has, for much of the post-war period, been characterised by labour shortages. The shortfall was at first made up by the inflow of skilled workers from East Germany, and lately by an influx of workers from neighbouring European and Mediterranean countries—the "guest workers". This chronic shortage provided an obvious incentive for German industry to retain by generous improvements in pay, training opportunities, status and promotion, a skilled core of native employees. Britain has enjoyed less flexibility with its labour force, since the workers it has imported from the Commonwealth have had residential rights, unlike the German "guest workers", and could not therefore be regarded as a reserve pool of labour to be drawn from or dispensed with as required. And at the same time, each upturn of the business cycle over the last decade has left behind a larger total of unemployed

labour than the last. Thus, the general employment situation in Germany has clearly helped to encourage a sense of obligation and commitment on the part of management towards its labour, which has, equally clearly, not developed in Britain. And management's attitudes have been reflected by labour in each country, with German workers taking a far more relaxed attitude towards impending change, and British workers a far more cautious attitude.

7. These general attitudes, shaped by the broad employment experience of the two countries, are reinforced by other factors which differentiate the two systems. The first of these is the institutions which have been developed to handle, among other things, the social aspects of change. We will return to these in greater detail in the next section. Here we wish to emphasise only one aspect.

8. In any German enterprise there is likely to be only one recognised union, since the structure of the trade union movement is along industrial lines. Similarly, there will only be one form of employee representation within the enterprise—the works council, elected by general franchise. Where change requires the redeployment of labour from one type of work to another, no institutional conflict therefore arises; for example, the members of one union cannot gain or lose jobs against the interests of members of another union. In Britain, on the other hand, the institutional arrangements are highly variable. Representation of employees is almost invariably through trade unions. Where no unions exist, works councils may be established voluntarily, though they tend to confine themselves to minor matters. Most typically, more than one union will be recognised in the enterprise, and there may be as many as two dozen. Such multi-unionism adds an extra dimension to the problems of change. One large group of craft unions exists to represent the interests of workers with particular skills, such as electricians or printing compositors, and they are naturally more reluctant to accept the dismissal of their members than they are to accept a loss of jobs as such. Where there is a surplus of labour, competition can sometimes break out between craft unions to preserve the jobs of their own members, even at the expense of the jobs of members of other unions. But even where the nature of the

change opens up opportunities for more jobs in a particular type of skill, it is unlikely that other employees of the same enterprise will readily be offered a chance to take them since they will not be allowed to gain the qualifications which are a condition of entry to the relevant union, and which can only be acquired through an apprenticeship governed by that union's rules. For workers in both countries there are obvious problems in acquiring new skills, especially for older people, but the essential element of flexibility that make redeployment within the enterprise on a large scale possible, and therefore reduces the need for dismissals, is often lacking in British industry. It is against this background that the vital necessity for government-backed industrial training schemes must be viewed. A measure of the scale of resources being devoted to training and retraining is that before the special measures to deal with youth unemployment in the summer of 1977 were announced, a million people a year were being trained or retrained. The externalisation of labour markets may be gauged, however, by the fact that some ten to 12 million job changes are recorded each year.

III

THE INSTITUTIONAL AND LEGAL FRAMEWORK

9. The second factor which differentiates the two national systems is the legal and institutional framework which has prevailed. In Germany, post-war legislation has concentrated heavily on establishing and building up enterprise level cooperative institutions, i.e. the works councils, and progressively enlarging their spheres of activity, whilst retaining the autonomy of collective bargaining. So when the aim of government policy has been to improve employee rights, the means has tended to be through these institutions. For example, in getting to grips with the problems of change, the law has been used to widen the areas over which joint decisions have to be reached, leaving crucial decisions on dismissals, transfers and redeployment to be settled in accordance with the conditions existing within the enterprise by the management and the works council—with a safety net appeals procedure to a labour court if the two sides cannot agree.

10. The British experience has been different. Reform by way of the institutions has been tried, but it has not so far proved effective or durable. Instead, government has laid down minimum legal standards which employers must observe in their relations with their employees, regardless of the institutional arrangements, although, of course, there is nothing to prevent unions and managements improving on these minima in collective bargaining. Some of this legislation has been directed at improving the conditions under which employees have to work, laying down, for instance, health and safety standards. But the main thrust has been towards regulating and improving the rights of workers at the point at

which their jobs are at risk—e.g. minimum notice of dismissal, compensation for redundancy, an appeal mechanism for alleged unfair dismissal, paid time-off to look for new work, and the assistance of government agencies in job placement. It is not altogether surprising that management and employee representatives should focus their attention on the dismissal option, since public policy points so emphatically in that direction, with state financial help to speed the departing worker on his way.

11. Nothing could be in starker contrast than the machinery in use in the two countries for dealing with the social aspects of change. In Britain virtually all consultation and negotiation over change takes place between management and trade unions, which have developed to the point at which they form the single channel for employee representation on all matters from national economic policy to the minutiae of day-to-day shop-floor problems; but in Germany a dual system exists for employee representation, with powerful trade unions operating largely outside the place of work and legally established works councils providing an enterprise-based negotiating partner for management on most matters affecting the enterprise. As a result, responsibility for resolving problems associated with change in Britain devolve onto trade union officials, at every level down to the lay shop steward, with their traditional role of vigorously defending membership interests; but in Germany, comparable responsibility usually falls on the works councils, which may not call strikes and are required by law to have regard for the interests of the enterprise as a whole.

12. The system in each country is the product of national history and experience. The German trade union movement was totally reconstituted after the Second World War into sixteen major industrial unions (although one important union representing white-collar workers, the DAG, was created on a trans-industry basis). In a series of legislative acts, works councils were created to represent the labour force wherever more than five people were employed, and in addition, rights of employee representation on the supervisory boards of large companies were introduced (varying from parity representation with shareholders in coal and steel, to

one-third representation in other industries). The functions of the works council are to police the operation of collective bargains made outside the enterprise and of laws designed to protect workers (e.g. health and safety) as well as to negotiate on any other matters that arise, such as the changes that are the subject of this report. The works council was the decisive instrument of representation in most of the case studies, with the worker representatives on the supervisory board playing at most a marginal role. The works council therefore appears to operate as an effective instrument of industrial change. There is, however, a tendency for councils to be dominated by representatives of the core labour force—skilled, German, and male—at the expense of fringe groups such as foreign workers, women and the unskilled. Given the dual system of representation, which virtually excludes trade union activity at the place of work, the works council can fail completely to represent the interests of any group of workers where it is incompetent, indolent, or under the domination of the employer.

13. In Britain, on the other hand, the trade union movement had known well over 100 years of continuous and unbroken development leading to a highly diverse pattern of unions—part industrial, part craft and part general. Their growth had been shaped neither by the law nor by the national trade union centre, the Trades Union Congress, and no alternative or complementary system of representation had been imposed on industry. As a result, a highly competitive structure of trade unionism has emerged as the sole conduit between management and employees for the great majority of British workers. The case studies reveal a very variegated pattern of representation, a variety which is characteristic of industry as a whole. In a sample of only seven companies, we found no common structure. In one, there was little or no union presence and only minor rights of consultation for the labour force through a poorly developed system of staff councils. In a second, a strong industrial union shared decision-making with management, who sought union assent to change, whether to a major strategic investment, or a comparatively minor production problem. In all the others, there was some form of multi-union representation, but even

here there were wide variations. In one instance, unions, while continuing their familiar role of negotiating wages and conditions, were learning to cooperate with, and even merge into, consultative committees separately established to consult with management over questions traditionally reserved to management alone, such as the company's investment programme. In another multi-union enteprise, inter-union rivalries were so great that it proved impossible to persuade the representatives of all the unions even to sit at the same meeting together. A further aspect of multi-unionism which emerged in one of the case studies was that management appeared to be somewhat confused as to which union, and at what particular level in the union, it should first approach in order to broach its plans for change.

14. For all these reasons—employment policies and the labour market background, the institutions which handle the social aspects of industrial change, and the direction of government legislation—employment has been more secure in Germany than in Britain. Employers had to value their labour more in the former than in the latter, if only because of its relative scarcity; at times of change, involving redeployment within an enterprise, the German worker can reasonably hope to transfer from one job or skill to another without risk of union objection, which hope the British worker cannot share. Further, the focus of government support in Germany has been on fostering mechanisms for resolving conflict peacefully within the enterprise; whereas in Britain support has in practice focussed on softening the impact of dismissal on the individual. The job security that these factors have between them engendered in Germany has given German workers a confidence with regard to change that appears to be lacking in Britain. One result of this is a greater willingness on the part of German management to reveal information about change to their employees far in advance of its implementation and, as we found in some of our case studies, beyond the requirements of the law—something that British managements are reluctant to do for fear of arousing opposition even before their plans are complete.

IV

LIVING WITH CHANGE

15. The contrasting institutional arrangements demand entirely different approaches to change on the part of British and German enterprises. In Britain, a company management has to determine its strategy for change very much according to the representational structure with which it has to deal. The result can be, at one extreme, decision-making by simple management fiat where, for example, no effective union exists to assert the rights of those employees who will be affected; at the other extreme, it may require the development of an elaborate set of quasi-political tactics for, first, enlisting the support of full-time union officials outside the enterprise, and then with their help, persuading the lower ranks of the union hierarchy—district officials or the company's own shop stewards—and, finally, the workforce itself to accept the proposed changes. In Britain, there are few general standards, other than those minimum requirements for discussions imposed by legislation, and rules exist only in enterprises where they have been learned through long experience of continuing change. Managements inexperienced in the process of change sometimes resort to specialised consultancies to devise procedures for them, and new systems and practices imported in this way can wither once their designers have left the enterprise to fend for itself.

16. Except in those increasingly rare cases where a workforce is not unionised, change of all kinds is handled by the unions as part of the collective bargaining process. Bargaining of this kind, as with wages and other material issues, carries with it the implicit risk of industrial action, which is the final sanction available to any union. In fact, open conflict

was almost absent from our case studies, but nevertheless the mere threat that it could happen heavily influenced the way some managements went about securing change. We have already referred to British management's reluctance to inform its employees at an early stage in the development of its plans for change. This is a factor which is hard to quantify. Even more difficult to pin down is the extent to which management anxieties over such aspects of change deter them from embarking on innovation at all, until circumstances force it upon them. This "apprehension factor" was clearly present in one of the case studies, where excessive manning levels were tolerated for many years until the business itself was in jeopardy.

17. The German system does not lack conflicts of interests, tensions between management and labour and opportunities for making bargains between the two sides. The difference from Britain is in the institutional structure within which these are resolved. As against the British system of representation by one or several unions—or by no union at all—the German worker is almost invariably represented by an enterprised-based works council.

18. If these works councils are always present, their behaviour is not always uniform. There is evidence that some adopt an abrasive attitude towards employers while other may over-identify their interests with those of the employer. The case studies suggest a middle way which we have designated "cooperative" conflict resolution". In this style a degree of sophistication has developed which enables problems to be solved while avoiding open confrontation. The system allows scope for management and workers to deploy sanctions against one another. The works council, for example, must give its consent to any dismissal, redundancy or redeployment of labour that management may think necessary. Its sanction is therefore to refuse to do so, or to delay giving its approval, or even just threaten delay. Management, for its part, may withdraw its cooperation, in such matters as the provision of information or material benefits such as cars or secretarial and other services to the works council. Again, a works council may agree to the dismissal of a particular employee whom management is

anxious to be rid of, and thereby earn a "credit" which it can look to management to redeem on another occasion when it wants a concession itself. In ways like these some of the bargaining ritual familiar in Britain between the shop steward and his management is reproduced in Germany. But, in general, the works council attempts the peaceful solution of disagreements within the enterprise, and it is a matter of policy to avoid, whenever possible, recourse to outside bodies like the labour courts or the unions. Thus, change is less likely to prove disruptive. Management can confidently expect the cooperation of its employees, to whom it is therefore ready to provide the information of impending change at an early opportunity. Both sides have learned to live with change as a continuing process, although as we pointed out in paragraph 12, where the works council does not fulfil its controlling function, change can have negative consequences for employees.

V

THE COSTS

19. In both Britain and the Federal Republic the economic and social costs of industrial change are borne—by no means in equal proportions—between the enterprise, the workers and the outside community. Labour legislation, in so far as it attempts to bestow job security, safe and healthy working conditions, training or retraining opportunities, obviously imposes costs on the enterprise, although social benefits and subsidies may relieve enterprises of part of those costs.

20. But legislation not only allows some of the financial cost of change to be transferred to the community—it also allows the transfer of people. Thus, British management's response to the demands of change, to dismiss "surplus" workers, involves a community cost in redundancy compensation, unemployment pay and, if unemployment continues for the individual, social security payments. If the redundant worker finds a place at a government industrial training centre and acquires a new skill, the cost again falls on the community (though some contribution is forthcoming from his former employer in the form of past payments to the State-run Redundancy Fund, the National Insurance Fund, and to the levy imposed in some industries to finance industrial training—the point being that the more readily the enterprise lays off rather than retains workers, the more value it receives for its payments under these various headings).

21. In Germany, the operation of stable employment policy on the part of large enterprises may not greatly swell the unemployment pool, and in-house training and retraining are a cost to the enterprise. But there is the "hidden" form of redundancy when older workers are retired early under

regulations that allow them to draw old age pensions from the age of 59. By exploiting its internal labour market, exemplified by the standard practice of halving recruitment to avoid dismissals, the German enterprise also transfers another social cost to the community, in the form of lost job opportunities for the unemployed, for school-leavers, and married women wishing to return to work. In addition, there is evidence of double standards in employment policy in some enterprises, with the "core" labour force—which, like the works councillors themselves, is predominantly German, skilled and male—being treated more favourably in terms of training and retraining, fringe benefits and stability of employment than members of fringe groups.

22. Change also inflicts a cost burden on the workers themselves. In Germany, stable employment policy is maintained at times of change and innovation by natural wastage, by a halt to recruitment, by early retirement, and by transfers within the plant. Redeployment can be cushioned by allowing workers transferred to a less skilled job to retain their former wage levels for a limited period, but it finally leads to a drop in income. A further cost to the community may be discerned in the case of such transfers that is not readily quantifiable, since they represent a deskilling process which eventually stretches right back to the numbers of young workers entering industry and the content of their training.

23. In Britain, as we pointed out earlier, the entry requirements of craft unions make it extremely difficult for the internal labour market of an enterprise to function in the same way as in Germany. Although transfers do take place, and, like the works councils, the unions will endeavour to avoid the workers having to accept lower pay, the main defence effort is to try to prevent dismissals occurring. A number of examples in the British case studies indicate a difference of purpose between the trade unions and their members in this respect, with workers displaying some eagerness to qualify for dismissal in order to collect redundancy compensation, even while being urged by the unions to fight against dismissal. And in one instance, where continuity of employment at another location a few miles away was offered by an enterprise to all the workers displaced by a

closure, only 88 accepted, while the remaining 482 opted for redundancy pay. It may be that the importance attached by union leaders and politicians to the ideal of full employment, even at a time of high unemployment in both countries, has made little impact on the behaviour of workers themselves: in Germany because of the workings of self-sufficient internal labour markets in big enterprises, with the security implicit in the stable employment policy; and in Britain because the fear of becoming unemployed is overridden by the irresistible lure of redundancy compensation, which may be the only opportunity most workers ever receive of acquiring a substantial lump-sum payment. In both instances a heavy part of both the immediate and the consequential cost is transferred to the community.

VI

SUMMING UP FOR THE FUTURE

24. The component factors of any industrial society are mutually interdependent. As society becomes increasingly complex and sophisticated it becomes more and more difficult to seperate the component factors into such simple categories as those of capital and labour, or to analyse the factors as isolated phenomena.

25. We have found this to be so in our studies. In identifying what we have called the adversary style of conflict resolution as practised in Britain and the cooperative style practised in the Federal Republic, we are conscious that these descriptions are themselves an encapsulation of many different factors. Among these are the broad political and economic experience of the two countries, the employment policies of enterprises, the general state of the labour market (prolonged demand plus a readily adjustable immigrant workforce in Germany, increasingly stagnant demand plus an unadjustable immigrant population in Britain), the framework of labour legislation, the organisations of management and labour, and many more. Furthermore, it is difficult to distinguish cause and effect. We might be describing the German style as adversary and the British as cooperative if the two countries had had each other's record of economic growth. Economic growth may be the great peacemaker. But of course even to envisage the two countries exchanging growth records would, in this context, be tantamount to assigning no importance at all to industrial relations as a factor in growth. It may well be that it is precisely the style of conflict resolution practised in the two

countries that does much to account for their economic records.

26. While recognising the intrinsic difficulty of disentangling the components of an economic system, we believe that in our study of change at enterprise level we have identified two themes of considerable significance which are well worth further thought and study. One theme concerns the competence of existing institutions to cope with future change; the other concerns the costs and benefits of enterprise employment policy.

27. In examining the whole range of factors involved in the successful prosecution of change in the two countries we were especially struck by the contrast between the institutional arrangements that prevailed in each. The contrast was between German uniformity and British variety. Whereas in Germany the management of a large enterprise could expect as a matter of course to deal with one union on questions of general interest to the industry, and with a well-established works council on questions of concern only to the enterprise, no such assured arrangements confronted British management. A British manager might have to deal with one union, or many, or no union at all; he might have to deal with union officials at national, district or local level; and he might have a system of consultation carefully built up over many years, or, lacking that, he might have to invent procedures on the spot to see him through a crisis.

28. The point may be put another way. In both countries the thrust of public policy has been in many respects in the same direction—towards enhancing the rights of workers to such things as consultation, better safety standards, greater employment and income security, improved training, and so forth. The contrast is the means by which policy is converted into practice. In Germany the development of policy has been accompanied by the development of institutional arrangements, notably through legally backed worker representation at the board and plant levels. In Britain the development of policy has not been accompanied by any comparable evolution of the institutions.

29. It has been put to us that the British system—or lack of system—has its advantages. The variety of procedures, on this argument, gives the British manager a flexibility denied his German counterpart. The results of our investigation do not support this view. Because institutions for consultation tended to be poorly established, consultation was haphazard and irregular, not a familiar part of the industrial environment. As a result, management approached change hesitantly, secretively and fearfully, while the workforce, as might be expected under the circumstances, responded suspiciously and aggressively. Change thus becomes inseperable from a mood of crisis. It is true that our research threw up a case where in Britain a company was so successful in promoting a rapid rundown in its labour force that a similar German enterprise (not covered by our case studies), faced with similar problems, sought its advice. But the British company, being virtually non-unionised, was of a kind increasingly rare among large companies, and we therefore do not regard it as a persuasive example of the benefits of a lack of uniform institutions.

30. The union as the major, if not the only, channel of worker representation seems now so well set a tradition in Britain that its modification by a German-style system of works councils legally detached from the unions seems to us an unlikely event, even if it were desirable. But since we began our study, an initiative has appeared which would be well worth consideration as a means of providing the uniformity of institutions that Britain now lacks, without breaking from tradition. This is the Joint Representation Committee proposed by the Bullock inquiry into industrial democracy. In the context of that inquiry the JRC was seen as a way of involving all the unions in an enterprise in the process of choosing, fairly and democratically, the worker-directors who would sit on the enterprise board. Although the ultimate fate of the Bullock proposals is still in doubt, the JRC at least seems a promising device. Future rights to consultation and joint decision-taking could be legally devolved on it. Recent legislation has provided new rights to individual workers (e.g. to periods of notice in the event of dismissal) and to unions (e.g. to warning periods in the event

of major redundancies). The provision of future rights might be made conditional on the establishment of a JRC. For instance, the committee could be given rights of joint decision on promotion, training and welfare, such rights not being available to individuals or unions not represented through the committee. Such rights could be gradually expanded until the incentive to form a JRC became irresistible. At the same time the JRC would have to accept certain obligations, perhaps, for example, the obligation to exhaust agreed disputes procedures before resorting to industrial action. This is only the sketch of an idea which would need much refining and discussion before it was introduced. But we believe that a development along those lines could help provide large British enterprises with the institutional uniformity they now lack; while it would be within the logic of recent legislative evolution, not against it. (Small companies were not part of the subject of this report which was concerned with large enterprises having, for the most part, strong channsel of worker representation. The machinery of change in small companies, less likely to have such strong channels, would be a particularly useful subject of further study.)

31. While the institutional framework established in the Federal Republic emerged from our inquiries as broadly satisfactory, some disquiet was noted over the relations between German unions and works councils. Although works council members were preponderantly trade unionists, it was felt that there was danger of the unions and the councils drifting apart, each following its own institutional imperatives at the expense of the other. For example, the dominance of "core" workers in the works councils may have led to the interests of fringe groups being ignored at plant level, no matter what union policy and aspirations for those groups might be. This seems to be especially the case when programmes of manpower rationalisation are introduced. Wider programmes of training and support for the unions and works councils might be a means of bridging the gap between the two, and additional means of integrating policy also merit consideration. (Programmes of training and support would, of course, be indispensable for the new means of institutional representation we have suggested for Britain.)

32. The costs and benefits of enterprise employment policy were outstanding among the larger issues thrown up by our inquiries (and also illuminated by discussions at a seminar held at the offices of the Anglo-German Foundation in May 1977). Those issues concerned the prevalent and obstinate growth of unemployment, both general and structural, in manufacturing industry and among white-collar workers.

33. As has been pointed out, the main purpose of the changes studied in this report has invariably been the same: to cut costs by reducing workforces. This has been true whether in Britain or Germany; whether the reduction has been abrupt or prolonged; whether by negotiation or management fiat; or whether by dismissal or natural wastage. Our report has noted the tendency of British management to opt for dismissal and that of German management for a more gentle attrition. But to those outside the enterprise the net effect is much the same: a reduction in the total number of jobs available in the economy. Indeed the relative smoothness with which jobs were reduced in Germany may help to accelerate the overall process of job reduction. And since the "core" workforce is more likely to be protected by the enterprise at a time of job shrinkage the axe is all the more likely to fall on the heads of fringe workers—the unskilled, women, immigrants. (This is one case where the fragmented nature of British worker representation may have value. For while "core" workers dominate the German works council, British fringe workers often have their own unions designed to protect their special interests.)

34. While the benefits of industrial change involving manpower reductions go largely to the enterprise—including the surviving workforce—many of the costs fall on the community. We were made aware of strong feelings, not least among the German representatives at our seminar, that the enterprises themselves were not seen to be bearing their full share of the costs. Whether this is quite fair is a matter for argument. Clearly, a firm that can contrive to remain viable and efficient can contribute more to the community in output, jobs and tax contributions than one that cannot. In any case a detailed study of this question lies outside the scope of this inquiry. Nevertheless, we observe that this feeling that

employers are not contributing fully to the social costs of change is being more and more frequently turned into concrete demands and direct claims against employers. As employers shed labour, the total numbers of unemployed have been seen to rise; so the unions counter this by claiming reductions in overtime, shorter working hours and earlier retirement, and they insist this should not be accompanied by any reductions in the pay of their employed members. In the union view this is a means of reducing unemployment. It can also be seen as a way of making enterprises accept a larger share of the costs imposed on the community by their rationalisation programmes.

35. This is a phenomenon which enterprises in both countries need to be aware of. While Germany and Britain have taken steps to ensure that unemployment no longer spells the disastrous drop in living standard which it once did, neither of our societies has yet rendered unemployment of no account. The possession of a job remains essential to the sense of well-being of the vast majority of the people, however well-off they may be without one. We suggest, therefore, that enterprises should actively consider ways and means of enlarging job opportunities (as the British Steel Corporation has already done). We are not suggesting feather-bedding or make-work programmes which would be self-defeating. What we are suggesting is that enterprises adopt larger employment targets as they now do larger sales or investment targets; that, in other words, their sense of commitment to their "core" workforces be extended to the community beyond the factory gates. The workforces of both countries are their fundamental asset still, and it makes social and economic sense for enterprises to be seen to make a contribution to their maximum employment.